# China's Porcelain Capital

# China's Porcelain Capital

## The Rise, Fall and Reinvention of Ceramics in Jingdezhen

Maris Boyd Gillette

Bloomsbury Academic
An imprint of Bloomsbury Publishing Plc
BLOOMSBURY
LONDON • OXFORD • NEW YORK • NEW DELHI • SYDNEY

**Bloomsbury Academic**
An imprint of Bloomsbury Publishing Plc

50 Bedford Square, London, WC1B 3DP, UK
1385 Broadway, New York, NY 10018, USA
29 Earlsfort Terrace, Dublin 2, Ireland

**www.bloomsbury.com**

**BLOOMSBURY and the Diana logo are trademarks of Bloomsbury Publishing Plc**

First published 2016

Hardback edition reprinted 2016
Published in paperback 2018



**British Library Cataloguing-in-Publication Data**
A catalogue record for this book is available from the British Library.

ISBN: HB: 978-1-4742-5941-5
PB: 978-1-3500-4482-1
ePDF: 978-1-4742-5942-2
ePub: 978-1-4742-5943-9

**Library of Congress Cataloging-in-Publication Data**
A catalogue record for this book is available from the Library of Congress

Typeset by RefineCatch Limited, Bungay, Suffolk

# Contents

# Tables and Illustrations

## Tables



## Plates

Plate 5: A carved Shang dynasty urn from Henan Province, dating to the late thirteenth or the early twelfth century BCE.

Plate 6: This *doucai* or 'joined colours' chicken cup from the Chenghua period (1465–1487) of the Ming dynasty is one of only 17 still in existence.

Plate 7: This porcelain salt dish, decorated with overglaze enamels and gilt, was made in Jingdezhen around 1760 for export to Europe. Jingdezhen potters appear to have copied a figure modelled by Johann Friedrich Eberlein for Meissen Porcelain circa 1746.

Plate 8: François Philipp photographed this porcelain worker carrying a board of greenware at the Ancient Kiln Folk Customs Museum in 2013. This technology was ubiquitous through the late 1950s, and remained in use for decades longer at some state and collective factories.

Plate 9: Frank B. Lenz, a Christian missionary living in the provincial capital of Nanchang, published this 'bird's-eye view' of Jingdezhen in the *National Geographic Magazine* in November 1920. Lenz visited in 1919 or 1920 and was surprised to learn that the city had no electricity, telephones, or newspapers. In his article about 'the world's ancient porcelain center', Lenz wrote that 'long hours, poor food, no rest days, and unsanitary living conditions cause a great deal of dissatisfaction among the laborers'.

Plate 10: Almost everyone visible in this photograph from February 1953 is wearing a Mao suit, also called a Sun Yat-sen suit. The photograph shows Chinese Communist Party leader Mao Zedong and others visiting the tomb of Sun Yat-sen in Nanjing.

Plate 11: During the Great Leap Forward campaign, Chinese leaders hoped to match the industrial output of Western capitalist nations through grassroots mobilization. Officials encouraged citizens to create 'backyard furnaces' and smelt steel, as seen in this photograph from the Weixing Commune. In 1960 alone, 10,000 Jingdezhen porcelain workers were diverted from ceramics manufacturing to steel production.

Plate 12: Frank Cosentino of Boehm Porcelain admired the foot-pedalled glazing machines that he saw at the For the People Porcelain Factory in 1974. He described how the operator pushed her pedal to make liquid glaze shoot up from the tub and coat the ware.
Plate 13: This employee at the Ancient Kiln Folk Customs Museum glazed bowls by dipping. He and several other members of the staff came out of retirement to demonstrate traditional production methods for visitors.
Plate 14: The People's Porcelain Factory, January 2009. This state-owned factory won numerous ceramics prizes during the 1970s and 1980s, and was visited by national leaders such as Deng Xiaoping, who was China's paramount leader from 1978 to 1992. The People's Porcelain Factory went bankrupt in 1996, a casualty of the central government's decision to privatize the state sector. In 2009, private entrepreneurs rented some sections of the factory to produce rice bowls, while other areas were abandoned.
Plate 15: After the state and collective porcelain enterprises collapsed, porcelain workers over 35 had a difficult time finding jobs. Older women with experience as surface decorators could apply transfers or decals for private entrepreneurs, as this woman is doing. The work was temporary and paid by the piece.
Plate 16: The International Porcelain Market, shown here in September 2004, was built on part of the East Wind Porcelain Factory site after the former state enterprise was sold in 2002. A private developer collaborated with city officials to build storefronts, many of which were completed by the end of 2003.
Plate 17: When I met 24-year-old Yang Yajun in 2005, he was a private entrepreneur who made high-quality replicas of Yuan dynasty porcelain. Yang built forms by hand-pressing clay into moulds that he made from plaster of Paris. In this photograph Yang is carving surface decorations on a greenware charger.

Plate 18: Ceramic artist Wu Yan painted this cherry-blossom tile in October 2005. Wu was one of a select group of cadres' children who learned handicraft techniques from older masters in the late 1970s. She left her job in a state porcelain factory to become a private porcelain entrepreneur in 1996. In the 2000s, Wu Yan switched from selling historic replicas to contemporary art.

Plate 19: Entrance to imperial kiln excavation site, October 2005. Municipal officials hosted a celebration to commemorate excavation of the imperial manufactory on 31 December 2003. Eight months later, facilities at the site were still very simple. By the autumn of 2005, the city government had added signage, a fancy gate, public art, and lighting.

Plate 20: This porcelain dragon, made from 40,000 porcelain rice bowls and cups, stood in the People's Square in 2012. City officials commissioned public art as part of their efforts to redevelop Jingdezhen and make the city attractive to tourists.

Plate 21: The Sanbao International Ceramic Art Institute, shown here in 2003, was the first place in Jingdezhen to offer private artistic residencies. Li Jianshen founded Sanbao as an artists' colony in 1998, designing the site as a 'living tradition' that evoked the lifeway of a Chinese scholar and provided a comfortable work environment for contemporary artists.

Plate 22: This plaque by Norwegian artist Ole Lislerud is entitled www.Fortune.com and measures 2 metres × 1 metre × 1 centimetre. Lislerud began making large plaques during a residency at the Sanbao International Ceramic Art Institute in 2002, and continued to experiment with the plaques for the next decade.

# Acknowledgements

What a pleasure to be able to thank the colleagues, friends, students, family, and organizations who supported me as I researched and wrote this book. This project began in the East Asian Art Department at the Philadelphia Museum of Art in 2003, when Dr Felice Fischer, the Luther W. Brady Curator of Japanese Art and Senior Curator of East Asian Art, encouraged me to curate an installation of Chinese porcelain made for Indian Muslims. Studying those wonderful objects, and working with Dr Fischer and Dr Adriana Proser (now the John H. Foster Curator for Traditional Asian Art at the Asia Society), made me curious about Jingdezhen's ceramists. Thank you both for starting this journey and sharing your expertise and friendship along the way.

From January through to April 2005, I had the good fortune to be a postdoctoral fellow at the Freer-Sackler Galleries of the Smithsonian Institution. I thank Louise Cort, Curator of Ceramics, and Jan Stuart, the Melvin R. Seiden Curator of Chinese Art, for introducing me to the rich material, archival, and library collections of the Smithsonian and providing expert tutelage, then and ever since. During the second half of 2005, I conducted research in Jingdezhen for five months, thanks to an American Research in the Humanities in China grant from the American Council of Learned Societies. I thank the many artists, workers, retirees, and officials who shared their experiences and perspectives with me during this and other visits. In particular, I am grateful to Gong Meihua, Yang Yajun, Liu Wenjuan, Jiang Huaqing, Ouyang Xiaosheng, Wu Yan, Liu Jun, Tang Ke, Zhou Chunlin, and Feng Shangsong. Your friendship sustained me and made this book possible.

I thank Haverford College for sponsoring several research trips to Jingdezhen via the Faculty Research Fund. I am grateful to Haverford colleagues Dr Laurie Kain Hart (now Professor of Anthropology at UCLA), Dr Paul Smith, and Dr Paul Jefferson (now emeritus), who were wonderful intellectual interlocutors throughout this project. I thank my former students Diana Tung and Patrick Lozada, who accompanied me to Jingdezhen and helped make *Broken Pots Broken Dreams* during the winter of 2008–2009. Kathleen McGee, Administrative Assistant for Anthropology and Sociology, supported me in every imaginable way, including tracking prices of Jingdezhen porcelain at auction, transcribing videotaped interviews, helping with logistical arrangements, and hundreds of other things. Thank you for your cheerful, can-do attitude – you have inspired me for almost 20 years.

I made significant headway on this project as a European Institute for Advanced Studies Fellow at the Swedish Collegium for Advanced Study (SCAS) in 2013–2014. Dr Björn Wittrock and his capable staff created a stimulating and productive environment for me and all the fellows. My SCAS compatriots, all of whom worked on topics far in time and space from Jingdezhen, raised new questions and helped me craft my story to reach a broad audience. I especially thank Petter Johansson, Pärtel Piirimäe, Marco Caboara, and Michael Yonan. You enriched my time in Uppsala in so many ways.

I am enormously grateful to Dr Stevan Harrell, Dr Carsten Holz, Dr Tik-sang Liu, Dr Eriberto P. Lozada, Dr James L. Watson, and Dr Rubie S. Watson for arguing about ideas, thinking through issues, sharing networks, and providing opportunities for me to present my research. You are wonderful colleagues and friends. I thank Dr Ruth Garfield for giving me personal and intellectual support over many years. I thank Dr Neil Brownsword for inviting me to speak on Jingdezhen at the British Ceramics Biennial and the Ashmolean Museum, and Dr Barbro Klein and Dr Inga-Lill Aronsson for discussing heritage and material culture with me. I am grateful to Jennifer Ko, who helped me learn the specialized vocabulary of Jingdezhen's porcelain industry and read several Chinese sources. I thank filmmaker and author Margie Strosser for sharing her ideas about what makes a good story, and Louis Massiah, filmmaker and Executive Director of Scribe Media Center, for modelling how to create a story with utility. I thank colleagues in the History Department at the University of Missouri-St Louis, and Dr Priscilla Song, for reading and offering revisions to Chapter Four. I am grateful to three anonymous reviewers for Bloomsbury Academic, who responded enthusiastically to this project and helped improve the manuscript. Diana Tung also read the entire manuscript and offered very helpful suggestions. I thank Senior Design Editor Rebecca Barden and Editorial Assistant Claire Constable for making the concluding stages of this project so smooth and rewarding.

I am so grateful to my family, who has lived this research with me. My siblings Adam Gillette and Emma Gillette, husband Mohsen Ghodsi, father Charles Cable Gillette, and mother Meredith Visnow Gillette have listened, laughed, argued, complained, distracted and supported me. Thank you for everything.

# Notes on Names and Transliterations

'Jingdezhen' is the pinyin transliteration of the name 景德镇, the city known as China's porcelain capital. This name has been written in English using many other transliteration systems. I use pinyin throughout, but readers will on occasion see other spellings such as Ching-te-chen, Kingtehchen, and so on.

Many but not all of the names of individuals in this book are pseudonyms. Some ceramists specifically requested that I use their names as they hoped that my book might help their businesses. I use the names of some ceramists and cadres because they have appeared in publications. Readers who would like to purchase porcelain made by the ceramists whose experiences I present here should contact me at maris.gillette@yahoo.com.

# 1

# The World's Most Famous Ceramics and the People Who Made Them

Jingdezhen. For many, the name is synonymous with fine porcelain. The Louvre, the British Museum, the Smithsonian, the Vatican Museums, the Hermitage, the Topkapi Saray, and many, many other institutions around the world have porcelain from Jingdezhen, China. Art historians describe its qualities and significance with superlatives, noting its formal perfection, refined decoration, and lasting influence on world ceramics. Collectors pay astronomical sums for Jingdezhen porcelain: for example, on 10 April 2014, CNN reported that Sotheby's sold a fifteenth-century 'chicken cup', a tiny bowl painted with two roosters and a chicken tending her young, for $36.05 million USD.[1] The world's best-known ceramics manufacturers, including Wedgwood, Sevres, Meissen, Rörstrand, and others, were founded to compete with Jingdezhen. Numerous contemporary artists, of whom Ai Weiwei may be the best known, take inspiration from Jingdezhen porcelain, and travel to the city to create works.

And what is this porcelain? Jingdezhen ceramists have produced an incredible variety of wares over their long history of manufacture. The icy-blue glazed celadons called *qingbai* or *yingqing* – thinly potted ceramics with a hard white body incised and impressed with naturalistic designs and covered by a translucent glaze – were Jingdezhen's first porcelain to achieve international renown. From the tenth through to the fourteenth centuries, *qingbai* were enormously popular in China and traded widely throughout Asia (see Plate 1). Numerous kiln sites across south China tried to replicate Jingdezhen's celadons; according to ceramics scholars Rose Kerr and Nigel Wood, 'few wares in China's ceramic history have been copied more frequently and over a broader area than Ching-te-chen [Jingdezhen] chhing-pai [*qingbai*] ware'.[2]

Better known in Europe and North America is blue and white porcelain (*qinghuaci*), a striking innovation in world ceramics history that Jingdezhen ceramists began making in the early fourteenth century. The first blue and white wares were larger, stronger forms than the celadons, and painted with surface designs in cobalt pigment under a translucent glaze (see Plate 2). Near-Eastern scholar John Carswell suggests that Persian merchants played a role in the creation of this new commodity, by bringing to China the cobalt used for the surface design, and supplying the Persian metalwork and glass that influenced the shape and decoration of the wares.[3] Over four centuries, blue and white porcelain captured an enormous, transnational market, becoming the most widely traded ware in world history. Jingdezhen blue and white stimulated numerous imitations, in China, Vietnam, Korea, Japan, Turkey, Persia, Egypt, and Syria.[4] The potters of Delft made the best-known European blue and white ceramics, but ceramists in Italy, England, Germany and elsewhere also produced Jingdezhen-inspired blue and white wares.

Many other innovations accompanied these two breakthroughs. Jingdezhen ceramists used other pigments for underglaze decorations, and painted coloured enamels over the glaze, adding an additional firing at lower temperatures to harden the surfaces. They created numerous glazes, deep monochromes in yellow, red, and purple, and multicoloured flambé and peach-bloom. They trailed white slip (liquid clay) on the surface of pots, and glazed between the raised lines with turquoise, purple, and yellow to create porcelain like cloisonné (*fahuaci*). They made wares that resembled lacquer, wood, stone, jade, and fruit and vegetables. They replicated pots from kiln sites in other parts of China and wares made by their predecessors. Jingdezhen ceramists produced porcelain for all sorts of Chinese tastes, and for international consumers who craved distinctly different forms and surface decoration. From the simplest to the most baroque designs, from the tiniest and most delicate to the sturdiest and most vigorous forms, wares for use at the table and in industry, for collection and construction – Jingdezhen producers have made them all. Jingdezhen has played such an important role in world trade that some believe the town gave China its English-language name. Prior to 1004, Jingdezhen was called *Changnanzhen*, the town south of the Chang River. A few ceramists told me that if you say 'Changnan' over and over, it sounds like the English word 'China'.[5] Others remarked that in English, both porcelain and the country of its origin are called 'china', which they took as a sign of Jingdezhen's significance.

Collectors, connoisseurs, and historians have written in multiple languages about Jingdezhen's porcelain for more than 500 years. For 300 years, a smaller group of government officials, industrial spies, and aficionados have described the technical processes by which Jingdezhen porcelain is made, often with illustrations. Yet when it comes to the people who throw, press, and decorate these renowned and treasured wares, little has been published – even though Jingdezhen stands out from other ceramics production sites for the sheer number of people who work in porcelain manufacturing. Fang Lili, an ethnographer, historian, and artist born in Jingdezhen, wrote the two most important books about the city's ceramists. Her Chinese-language texts investigate nineteenth- and late twentieth-century potters, painters, kiln workers, seconds-sellers, and others who participated in making the city's prized commodity.[6] In English we have no full-length studies of Jingdezhen's ceramists. Shorter publications are scattered in journals, catalogues, trade magazines, and chapters of edited volumes.[7]

In this book, I write about the artists, artisans, and workers who made Jingdezhen's porcelain, focusing on the period from the end of the Qing Dynasty (1644–1911) through the first decade of the twenty-first century. I investigate the conditions under which ceramists could make a good living from porcelain manufacturing, and the circumstances which made that difficult. A key player in ceramists' success or failure, happiness or misery, has been the government. How the Chinese state consumed, invested in, taxed, and managed Jingdezhen porcelain has profoundly shaped ceramists' lives, the local environment, and the quality, quantity, and types of wares produced.

Jingdezhen's story will fascinate those who are interested in ceramics or Asian art. Jingdezhen came to be the porcelain capital because of its unique system of manufacture that enabled ceramists to produce large quantities of exquisite porcelain without mechanical assistance. This system arose from conjoining state intervention and local social formations. For students of modern Chinese history, Jingdezhen's trajectory adds depth and dimension to the national narrative, and problematizes what we think we know about China. Many scholars regard Chinese Communist Party rule as a radical break with China's imperial past, and indeed, the Chinese Communist Party put an end to many long-standing social practices and cultural beliefs. Yet when it comes to porcelain manufacture, the Party under Mao Zedong built on and extended key imperial policies rather than rejecting them. Jingdezhen's most radical break with history occurred when the government decided to privatize the porcelain industry in the late twentieth century. Readers interested in industry, labour, and capitalism may reconsider their understandings of economic growth from seeing how China's porcelain capital developed over time. While Jingdezhen's initial success in global trade had much to do with its monopoly commodity – true porcelain, made from naturally kaolinized china stone – big government was essential to manufacturing's expansion, the high quality of wares created, and a thriving local economy. After Jingdezhen became a massive ceramics producer, state management, investment, and consumption kept the industry strong. Only when the government withdrew from Jingdezhen, whether because of state failure, economic policy decisions, or both, did manufacturing decline, the economy shrink, and ceramists suffer. In this history lie lessons for those who want to understand the political and economic arrangements that undergird the health of manufacturers around the world.

## The basics: Clay, fuel, and transportation

Before Jingdezhen became a massive modern and pre-modern manufacturing complex, before the Song dynasty emperor Zhenzong set off the town's industrial development by admiring its icy-blue celadons, before local farmers began making ceramics during the agricultural slack season, was Jingdezhen's china stone. Today this distinctive resource draws ceramists from thousands of kilometres away to Jingdezhen. It differs not only from the materials that Europeans and Americans use to make porcelain, but also from the clays that north China potters employ.

Jingdezhen porcelain is not made from true clay. Clay usually comes from the weathering of volcanic rocks that are high in feldspar. By contrast, Jingdezhen porcelain is made from a rock that is high in mica. This stone, called china stone, petunse, or porcelain stone (*cishi*), produces ceramics that are extremely white, pure, and translucent. It naturally has low levels of clay (kaolin) mixed in, but only between 10 and 20 per cent. China stone can be found in south China, Korea, and Japan – and nowhere else.[8] It is abundant in Jingdezhen. Even now the local government estimates that Jingdezhen has at least 300 more years of china stone to mine.[9]

Today's ceramists use approximately the same methods to turn china stone into a clay-like material as ceramists did in the eleventh century. The first step is to mine the china stone, either from an open quarry or underground. The next is to pulverize it. Jingdezhen's residents have long harnessed the area's many streams to power mallets that pound china stone to a fine powder. In the early twenty-first century, visitors could still see this type of water wheel and crusher at work (see Plate 3). Government officials pushed ceramics producers to switch to mechanized crushers in 1955, but the practice of using water-powered mallets never stopped completely.[10] After the stone is pulverized, the clay maker levigates it in water, to separate the coarse from the fine material (see Plate 4). Then s/he ladles the muddy stone out of the water and lets it dry and stiffen. Workers use a simple wooden form to shape the material into small white bricks or *bai dunzi*.[11]

This phrase is the basis for the English word 'petunse', a synonym for china stone. Potters buy the bricks and mix the china stone with water to make the equivalent of clay. Processing is arduous, particularly if one does it manually. In 2005, I apprenticed to a potter who made replicas of fourteenth and fifteenth-century blue and white porcelain. We stamped the muddy china stone with our feet to distribute the water and remove air – the same method potters used in the fourteenth century, and probably earlier.

Prepared china stone is wonderful and difficult as a medium for forming. My master instructed me in the local method, which is to throw thick forms off the hump and trim extensively after the pot dries, when it is brittle and easy to break. China stone 'clay' is very soft and pliable, and shapes easily on a potter's wheel, but it is not very strong. *Qingbai* was made from this single stone, and potters threw small pieces. In the late thirteenth century, ceramists began adding kaolin, a true clay, to the mix. This allowed them to make the large chargers, dishes, urns, and vases that we associate with the blue and white porcelain of the Yuan or Mongol dynasty (1260–1368).

The word 'kaolin' comes from the Chinese phrase for 'high ridge' (*gaoling*), which is the name of a range of hills near Jingdezhen. Gaoling has true clay, and at various times over Jingdezhen's history ceramists have mixed more or less kaolin with china stone to make their medium. During the seventeenth century, potters added the largest proportion of kaolin to porcelain stone: 50 per cent.[12] Since the nineteenth century, potters have been using a mix of 30 per cent kaolin and 70 per cent china stone – unless they are making historic replicas, or counterfeit antiques, in which case they use recipes that produce a better match with the period from which they are copying (see Chapter Six for more on antique replicas). The original Gaoling mine was used heavily from the seventeenth through to the nineteenth centuries, and finally closed in 1969.

Jingdezhen could not have become the porcelain capital (*cidu*) without china stone, the key ingredient for the area's ceramic bodies and glazes. But transforming china stone into porcelain, with its glassy, hard body, requires firing, and firing requires fuel. From the eleventh century until the early 1950s, local ceramists fired in kilns that burned pine logs or brushwood. The best ceramics came from wood-burning kilns; brushwood was used for coarse wares. Initially the wood came from abundant pine forests around town. But as production expanded, Jingdezhen became severely deforested. Fuel was hauled in, or floated down rivers and streams. The high cost of transporting pine logs from farther and farther away led the government to switch the city to coal-fired kilns in 1956.[13] Coal had been used for centuries in north China kilns, also because of deforestation, but never in the south. Coal was Jingdezhen's dominant fuel until 1992, when the industry began switching to propane gas. Municipal officials ordered Jingdezhen's last coal kiln to close when I was visiting there in September 2004.

Another important element for Jingdezhen's success was the ability to transport porcelain to markets in China and beyond. Even now, Jingdezhen is a small city by Chinese standards, located in a poor rural province far from the provincial capital, let alone such major metropolises as Shanghai, Beijing, or Canton. The Chang River, which until the mid-twentieth century formed the western boundary of the city, was key for getting porcelain to markets. Transporters could sail boats – or when the water was shallow, pull them with ropes from the water or the shore – down the Chang River to Boyang Lake, which joins the Yangzi River. From there, Jingdezhen porcelain went north and east to Nanjing and Shanghai. Shanghai shippers took the porcelain on the Grand Canal to Beijing and the imperial court. This route enabled Jingdezhen to produce wares for successive emperors, which in turn led to the massive expansion that made the city porcelain manufacturer for the world. For markets to the south, porcelain was carried out on the backs of porters who trekked it over hilly paths to Canton. From Canton it shipped around the globe.

## Research materials, methods, and scope

I draw on historic, economic, and ethnographic sources to write the history of the porcelain capital and its ceramists. Very few records exist about Jingdezhen's early porcelain producers, and the primary sources are written in literary Chinese that is difficult for the non-specialist to read. I use secondary scholarship, in Chinese and English, by historians, art historians, archaeologists, ceramics scholars, and curators to describe Jingdezhen from the eleventh through to the early nineteenth centuries. I also rely on translations of primary source documents, many of which were published by Robert Tichane in his compendium *Ching-te-chen: Views of a Porcelain City*.[14] My research also included studying porcelain from this time: the pots themselves reveal glimpses of the labour that produced them. The narrative that we can glean about Jingdezhen and its potters, particularly during the early periods, is necessarily partial. The observers who created the few accounts that we have spoke from their own perspectives, and it is difficult to assess their reliability. Because of this, my analysis in Chapter Two is somewhat speculative.

When we get to the nineteenth and twentieth centuries, sources are more abundant and varied. Beginning in the nineteenth century, numerous Chinese elites and foreigners visited Jingdezhen and published accounts of what they saw. I read these primary sources in Chinese and English. I studied government documents printed from the 1950s through the early twenty-first century. These included reports for official consumption, public records such as yearbooks and gazetteers, oral histories collected by government work teams, state-sanctioned histories, and newspaper articles published in state media. Like all historic documents, these government materials have their ideological biases, but exacerbated in this case by state control over the press. By using government documents from periods when the 'party line' was quite different (e.g., the 1950s versus the 1990s), I am able to correct for some of this bias. Another limitation of these sources is that they privilege state policy over ceramists' experiences. From time to time, official documents do include records of how ceramists responded to political campaigns and technological changes, but rather infrequently. To understand how locals felt about working in porcelain, I supplement the printed records with life histories that I gathered during my field research. This too leads to biases: I have much richer characterizations of more recent decades, and my own participant-observation from 2003 to 2010, and fewer accounts of the 1950s and 1960s. Additionally, memory is tendentious, shaped as much by present circumstances as by what happened in the past. I indicate in the text when the sentiments expressed are remembered years after the events discussed. My other research methods included examining photographs of Jingdezhen from the 1920s through to the present, viewing films from the 1980s, and scrutinizing many pots.

For the contemporary period, ethnography was my main research method. I first visited Jingdezhen in the winter of 2003–2004, when local officials, ceramics entrepreneurs, and media professionals were preparing for Jingdezhen's 'one thousand year anniversary' as the porcelain capital in 2004. Ironically, 2004 was also the year that the central government took the title of 'porcelain capital' away from Jingdezhen, awarding it to the city of Chaozhou in Guangdong (Canton) Province. Between 2003 and 2010, I travelled to Jingdezhen seven times, living there for periods as short as three weeks and as long as five months (from August through to December 2005). I spent time with residents in their workshops, factories, homes, and public spaces, learning as much as I could about how people made a living in porcelain. I met throwers, press-moulders, slip-casters, underglaze and overglaze painters, glazers, gophers, apprentices, mould-makers, greenware carriers, porcelain inspectors, kiln loaders, decorators who applied transfers to unfired porcelain bodies, shopkeepers who sold finished porcelain or supplies for ceramists, brush-makers, box-makers, studio artists, college and university professors of ceramics and ceramic history, government officials in tourism, education, porcelain, and taxation, laid-off ceramics workers doing temporary jobs or looking for work, and industry retirees. I conducted formal interviews and had many informal conversations. I watched people work and I worked myself whenever I could.

For five months in 2005, I apprenticed to an antique-replica maker who ran a small workshop with his wife and two cousins. My apprenticeship gave me first-hand knowledge of the skills needed to make Jingdezhen porcelain, and the crucial role that repetition plays in developing them. In Jingdezhen's ceramic industry, labour is extremely specialized. Unlike studio potters, Jingdezhen ceramists do only one small part of the production process, which makes working in porcelain very monotonous. My experiences as an apprentice helped me understand why so many people in Jingdezhen talked about their work in porcelain as simply 'a way to make a living' (*kao zheige shenghuo*), rather than as an artistic endeavour or a vocation. More sporadically, I worked in factory settings. I did simple tasks like affixing bar codes to blue and white rice bowls for export, or wrapping pots in paper packaging. Another activity I learned was how to play mahjong (at a beginner level). When work is slack, many porcelain workers gamble for small stakes, and many laid-off ceramists and retirees gamble at mahjong, as well as cards and chess (*weiqi*).

My primary consultants were ceramists who learned to work in porcelain by apprenticing in workshops and factories. As was characteristic of the city's population, they had little formal education. The majority of Jingdezhen's residents had completed their education with primary or middle school. A smaller group of consultants in their 30s and younger had completed high school. Many of the elderly residents I met had almost no formal education, and a few were illiterate. Almost all of the city's schools were built after 1950. Today Jingdezhen has several universities, colleges, and research institutes, of which the biggest and best known is the Jingdezhen Ceramics Institute (a research university). I met and interviewed professors, artists, students, and officials who worked at these places. But in large part, my understanding of Jingdezhen is based on the experiences of those who learned ceramics on the job, rather than in the classroom.

I begin this story of the porcelain capital before the advent of writing, but my focus is Jingdezhen during the last 150 years. I draw on anthropological and historical research on Chinese family and society to delineate the lives of ceramics workers in the earliest periods. As we enter periods when textual, visual, and ethnographic source materials are more abundant, I create more detailed and colourful portraits. The fullest narrative comes from the second half of the twentieth century and first decade of the twenty-first.

## Jingdezhen's longue durée

As I watched glowing flames from the peek holes in a kiln door on a cold day in January 2004, the man tending the firing spoke to me. 'See this?' he asked, gesturing at the two holes and the small wooden door below them. 'What does it look like? It's a woman's body,' he said, not waiting for an answer. 'Look, you can see her figure, like this. The doors of Jingdezhen's wood-burning kilns always look this way. Do you know why?'

> Once, long ago, the emperor ordered the workers in Jingdezhen to make him a set of gigantic urns decorated with dragons. The potters tried and tried, but the urns kept failing during firing. Over and over when they opened the kiln the urns came out misshapen. As the numbers of failures increased, the officials in charge grew angry, and threatened the ceramists with severe punishment, even death, if they didn't complete the urns. So the next time that the urns were firing, the kiln master's daughter, a young girl, snuck into the kiln shed and leapt into the kiln, burning herself to death in its flames. When the firing finished and the unloaders opened the door, the urns had fired perfectly. When we build wood-burning kilns in Jingdezhen, we always make them with the symbol of her body, like this, to remember her sacrifice.

The man who first told this story to me was the kiln master (*bazhuang*) at a wood-burning kiln located at the Ancient Kiln Folk Customs Museum, a combined living-history museum and production site that government officials constructed from salvaged portions of traditional workshops during the late 1970s. Over the course of my research in Jingdezhen, I heard this tale at least a dozen times. Sometimes the person who leapt into the kiln was a male potter. This story appears often in written records about Jingdezhen, usually as a history of the Ceramics God, also called the God of Wind and Fire. The oldest written account from Jingdezhen dates from the late seventeenth century, but Rose Kerr and Nigel Wood have found a related story that was written in the mid-fourth century.[15]

While I dutifully wrote this story down in my notebook each time that someone recounted it, only later did I realize that the tale communicates essential information about working in porcelain in Jingdezhen. The story tells us that porcelain is an uncertain and difficult way to earn a livelihood, potters must make physical sacrifices if they want to succeed, and the Chinese state has significantly influenced ceramists' lives and wellbeing. For most of the period from 1004 to 2010, the government invested in and supervised porcelain manufacturing, and consumed a portion of Jingdezhen's wares. From the Song dynasty through to the end of the eighteenth century, the overall trend was for government officials to play larger and larger roles in production. The exceptions were periods when the country was at war and the government failing. When the Chinese Communist Party came to power in 1949, they followed suit, investing massively in porcelain production, expanding what Jingdezhen made and managing every aspect of the industry. The Communist government directed and supported porcelain manufacturing for forty-six years, until officials privatized all state and collective enterprises in the mid-1990s. Since 1998, Jingdezhen's porcelain manufacturing has been private.

Jingdezhen's history of big government is crucial to understanding the views, sentiments, and actions of ceramists past and present, as well as the industry's economic health. Contemporary Jingdezhen has deindustrialized, and its porcelain lost ground relative to other Chinese ceramics manufacturers. In recent years, many Jingdezhen ceramists have shifted to making art porcelain with handicraft methods, especially antique replicas and ceramics influenced by the Chinese national school of painting (*guohua*). Local officials have invested in urban reconstruction and tourism, rather than directing ceramics manufacturing. We can only make sense of the impact of these changes if we understand the history of state involvement in Jingdezhen, and how political decisions led to this outcome. Shedding light on these developments, and on how ceramists have responded to them, is the promise of this book.

# 2

# Creating a Porcelain Capital, Prehistory to 1785

About 20,000 years ago, hunter-gatherers living in a cave 170 kilometres northeast of Jingdezhen made the oldest-known pottery in the world. Remnants of these vessels were found by archaeologists in the Cave of the Immortals (*Xianrendong*) near Jiujiang, Jiangxi Province. Someone flattened clay sheets, or coiled clay and beat it with paddles to smooth the sides, to make the pots. The containers were fired at low temperatures, probably in open fires. Scorching on the sherds suggests that they were used for cooking, although archaeologists have not discovered any food residue.[1] Some scholars, extrapolating from the social organization of contemporary hunter-gatherers, suggest that women made these first ceramic wares.

China produced the world's first high-fired ceramics about 18,000 years later. Some examples come from Wucheng, Jiangxi, about 300 kilometres southeast of Jingdezhen.[2] These good-quality, glazed, high-fired stonewares date to the Shang dynasty, 1600–1046 BCE. They are carved and incised, exhibiting the same detailed craftsmanship as the well-known bronzes from this period (see Plate 5). Archaeologists have uncovered dedicated sites for ceramics production near Shang palaces, suggesting specialized manufacture, perhaps even that ceramics was a full-time occupation.

Scholars believe that men made these high-fired wares. Shang society was patriarchal, agrarian, and stratified. People traced kinship through male links and men ruled the empire. From this time forward, male-centred genealogies, male rulers, and male potting characterize most of Chinese history.

Not until 1,000 years after the Shang dynasty do we find the earliest records written about ceramists. This is also when a term for high-fired ceramics, *ci*, appears in Chinese texts. In the Chinese language, ceramics are either low-fired *tao* or high-fired *ci*; there is no special word to differentiate porcelain from stoneware, as there is in English. Rulers during the early Han dynasty (206 BCE–220 CE) talked about potters when they were worried about food scarcity and too many people turning away from agriculture. For example, the Emperor Jing complained in 141 BCE that potters did 'non-essential things'.[3] Historians believe that people around Jingdezhen made ceramics during the Han dynasty, perhaps as early as 25 CE.[4] Records of taxes on ceramists and other craft producers exist from this period, as well as sumptuary regulations that tried to limit the use of luxury products. Sources reveal almost nothing about the people who made the pots.

The earliest known ceramist from Jingdezhen was a man named Tao Yu, who produced wares during the early Tang dynasty (618–906). He presented celadons to the court in 619 or 621.[5] The emperor apparently liked Tao's 'imitation jade', because the government set up an official bureau in the Jingdezhen area, presumably in relation to ceramics. Records from this time mention another Jingdezhen potter named Huo Zhongchu. Later documents speak of the Tao kiln and the Huo kiln, which suggests that each man headed, or was part of, a family or lineage that made ceramics. Tao's and Huo's pots were stoneware, not porcelain. The evidence we have suggests that Jingdezhen potters did not make porcelain until the late tenth century; the earliest known Jingdezhen porcelain was found in a Song dynasty tomb dated to 1000.[6]

If we take the early Tang dynasty as the start of Jingdezhen's ceramic production – still not yet porcelain – then Jingdezhen appears late in China's ceramic history. Potters in north and south China made earthenware and stoneware for millennia before much was happening in Jingdezhen. While none of these wares had the global scope of Jingdezhen's Song-dynasty *qingbai*, many travelled all over China, across the Asian continent and overseas. From 200 BCE onward, the Chinese exported pottery. During the seventh and eight centuries CE, international trade was big business in China. Northern white wares from Dingzhou, Yue celadons from south China, and the tri-colour splashed pottery for which the Tang dynasty is so well known attracted consumers in Asia, Africa, and the Middle East.[7]

## Early Jingdezhen

The literate elites whose records we rely on to understand late-imperial Chinese society wrote almost nothing about the lives of Jingdezhen's ceramists. Scholars surmise that until the late tenth or early eleventh century, farmers around Jingdezhen made pots when agricultural work was slack. Families had small ceramics workshops and fired their wares in dragon kilns, narrow single-chamber structures built up the area's low hills that had fire pits at the base and stoking holes at the sides.[8] For fuel, locals chopped wood from the thick forests. We don't know whether the men who formed *qingbai* celadons did all the other steps too: mining china stone, logging, building kilns, making glazes, and decorating, as well as farming. We do know that most of the area's kilns were outside of the town proper. Jingdezhen was small, with some local officials and a large rural population.

To envision how these potters lived, we have to extrapolate from social and cultural patterns that characterized other parts of China. From the tenth through to the nineteenth centuries, most of China's population was agrarian. Men headed families, relations between men created kinship (e.g., fathers, sons, and brothers were named in written genealogies), and men inherited the family property. Unlike Europe, China did not practise primogeniture: a man's property was divided equally among his sons. Women married out of their birth families into their husband's family, and typically shared living space with their husband's parents. Parents did not want their daughters to marry into families that shared their family name (traced through men), as it was believed that such families must be related. Relatives who shared a family name were likely to live in close proximity, so women often had to leave their home villages when they married.

Families venerated their male ancestors, as well as a number of gods and Buddhist deities, in a syncretic practice that scholars usually call popular religion. Jingdezhen and its surrounds still have numerous temples to Buddhist and Daoist deities, some of which, locals say, date from the Song dynasty (960–1279). In some families, a male ancestor demarcated a lineage, whose members owned property together, cooperated for defence, ran schools for their male children, and participated in collective rituals. Lineages existed in and around Jingdezhen. For example, the Wei lineage moved to the area during the Yuan dynasty (1260–1368), and monopolized the profession of kiln-building from the late seventeenth through to the nineteenth centuries.[9]

Written records and the oral testimony of residents confirm that Jingdezhen had lineage schools. Beginning in the Tang dynasty (618–906), the imperial government began holding examinations to recruit men into civil service, rather than awarding government office based on heredity. This gave families a chance for social mobility: a family's fortunes rose dramatically if one of their own tested into the government. During the Song and Yuan dynasties, Jingdezhen had an elite academy that produced scholars and officials.[10] In later centuries, the area's scholarly achievement declined noticeably, perhaps because of the ceramics industry's increasing dominance. Several locals told me that when the Chinese Communist Party took over the local government in 1949, only a tiny proportion of the population was literate. Officials had to recruit high school graduates from Shanghai to come and work for the city as administrators, researchers, and educators.

While our knowledge of Jingdezhen ceramists during the Song and Yuan dynasties is limited, we are better informed about elites who had connections to the town. By the opening of the eleventh century, Jingdezhen's icy-blue *qingbai* porcelain was renowned as 'the jade of Raozhou' (Raozhou was the name of the prefecture in which Jingdezhen was located). In 1004, *qingbai* caught the attention of the emperor of China, the man best known as the emperor Zhenzong (r. 997–1022), although he changed his reign name repeatedly over his long rule. Zhenzong was the third Song dynasty emperor, and many Chinese historians consider a treaty that he concluded in 1004 with the northern Khitan people to be the key event of his reign.[11] The Treaty of Shanyuan ended a long border war and brought the Song dynasty a century of peace, and initiated an even longer period of nationalist demands to recover lost territory and wreak revenge. From the perspective of world history, however, Zhenzong's decision to patronize a small ceramics-producing town south of the Chang River was even more significant.

The emperor Zhenzong renamed Jingdezhen, ordering that the town change its name from 'the town south of the Chang River' (*Changnanzhen*) to 'the town of [the] Jingde [emperor]' (*Jingdezhen*), after the reign name (Jingde) that he used from 1004 to 1007. He sent an official to take charge of supplying the court with tribute wares. This man, known as a 'supervisor of the market town' (*jianzhen*), also collected taxes on porcelain, controlled fuel for kilns, and kept law and order.[12] The official requisitioned porcelain from private producers and sent the wares to the emperor. Local gazetteers indicate that Jingdezhen had 300 workshops, run as family undertakings, at this time.[13] We perhaps catch a glimpse of Jingdezhen ceramists from a poem by the official Mei Yaochen (1002–1060), who was born in Anhui, a neighbouring province. Mei wrote in 1036:

> The potter
> Pots cover every inch of space before the door
> But there's not a single tile on the roof
> Whereas the mansions of those who wouldn't soil their fingers with clay
> Bear tiles overlapping tightly like the scales of a fish.[14]

In 1082, China's central government established an office to manage Jingdezhen's porcelain trade.[15] This contributed to the development of land routes for ceramics commerce; sea routes had been operating for centuries. Officials strictly monitored porcelain quality, and punished everyone in the supply line for any irregularities. Jingdezhen grew from a regional ceramics supplier into an international one. Archaeological evidence indicates that *qingbai* production reached a high point between 1075 and 1150.[16]

The Song government lost north China to the Khitan in 1127. While bad for the dynasty, the loss was good for Jingdezhen. Skilled potters from northern ceramics centres fled south when the imperial court moved to Hangzhou in Zhejiang Province. Northern ceramists brought their technologies, forms, and decorative styles to Jingdezhen, and the town's porcelain soon eclipsed the popularity of northern white wares.[17] Jingdezhen ceramists imitated wares from the northern kilns, particularly the celadons from Yaozhou in Shaanxi and white wares from the Dingzhou kilns in Hebei.

Jiang Qi was an official familiar with Jingdezhen who wrote the *Record of Ceramics* (*Taojiluo*), between 1214 and 1234. He provides tantalizing details about local porcelain producers.[18] Jiang writes that production was seasonal, because the clay was too stiff for shaping during the winter. Ceramists who produced for the court were given land to cultivate near the manufacturing sites, and were not paid regular wages. Production was specialized, with different individuals preparing clay, throwing pots, trimming, glazing, decorating, carrying greenware, loading it into saggars (protective cases for firing), and firing. Officials taxed kilns based on size, inspected porcelain quality, and regulated use of glazes. Jiang complains that they required constant bribes. The taxes from porcelain production paid for the salaries of officials, the police, and soldiers, and funded collective rituals, holiday observances, and the halls where men took the civil service examinations. Porcelain taxes also paid welfare to the widows and orphans of ceramists. Jiang writes that the industry's production depended on local farming: the porcelain workshops could not open if the harvests did not produce enough food.

When the Mongols took over China and founded the Yuan dynasty (1260–1368), the new government intensified its role in Jingdezhen porcelain manufacture. The Yuan central government set up a Porcelain Bureau (*ciju*) in Jingdezhen in 1278, increasing the number of officials who regulated production. Officials supervised the use of clay designated specifically for imperial porcelain, watched over the quality of glazes, and punished people who used grades of glaze that the state deemed inappropriate to the wares produced. Other officials governed the taxation of private ceramists. Still others were in charge of ensuring that a special white ware, inscribed 'imperial palace' (*shufu*), made it to the court.[19] To make *shufu*, officials conscripted private potters who, when the government did not requisition their services, made commercial wares. Jingdezhen had a professional ceramics workforce by the fourteenth century, as demonstrated by the specialized tools made for individual ceramists that archaeologists have found.[20]

The Yuan imperial household was descended from Chinggis Khan, and by the end of the thirteenth century, descendants of Chinggis Khan controlled most of the Eurasian land mass. Chinese traders from the southern city of Quanzhou brought cobalt from Persia to Jingdezhen, and encouraged ceramists to paint unfired porcelain surfaces with this blue pigment. Then they sold the blue and white (or underglaze blue) porcelain to Middle Eastern and other international consumers.[21] Under the Yuan government's management, Jingdezhen became a global manufacturing centre for blue and white porcelain. In the thirteenth and fourteenth centuries, Chinese elites thought blue and white porcelain was vulgar.[22] Jingdezhen's main market for underglaze blue ceramics was outside China.

Yuan emperors regularly appointed non-Chinese to manage Jingdezhen's Porcelain Bureau. The first supervisor was Nepalese, and his successors included Persians, Mongolians, and Muslims from Central Asia.[23] Historian Anne Gerritsen has mined the Song, Yuan, and Ming dynasty texts on Jingdezhen for evidence that the town had an international population, given that it produced a commodity that appealed to foreigners in many parts of the globe.[24] She finds almost nothing to suggest that Jingdezhen was shaped by global influences, beyond the fact that it was a busy production site. One of her sources is Hong Yanzu (1267–1329), a schoolmaster in Jingdezhen who later became an official. He described Jingdezhen as a busy commercial place, noisy, littered with pot sherds, and polluted by the smoke from wood-firing kilns. Hong wrote that local ceramists were skilled, but thought the town's residents lacked 'proper ceremony'.[25] This likely means that locals spent most of their time making and selling porcelain rather than following literary or ritual pursuits.

The key technologies for making ceramics, such as throwing, trimming, and moulding, were developed centuries before Jingdezhen appeared as a production site. What Jingdezhen contributed to the history of ceramics technology was organization. The city pioneered labour specialization, where an individual performed only one aspect of manufacturing in a kind of handicraft assembly-line process. Many ceramists worked on a single vessel en route to the finished product: as locals put it, '72 hands to finish a pot'.[26] This method meant that producers could create wares of high quality at a consistent rate. Jingdezhen ceramists became extraordinarily skilled at the tasks they performed – but they produced as specialists, not as studio potters. If Jingdezhen ever had a studio tradition, it was over by the fourteenth century. Making porcelain in Jingdezhen was and remains a collective endeavour.

## The imperial manufactory and rise of the porcelain capital

From 1352 to 1354, Jingdezhen saw heavy fighting as the Yuan dynasty fell. Porcelain production temporarily ceased. Ultimately, Zhu Yuanzhang, the son of a poor farmer from Anhui, founded a new empire that he called the Ming dynasty (1368–1644). In the second year of his reign as the Hongwu emperor (r. 1368–1398), he built an imperial manufactory on Pearl Hill (*Zhushan*) in Jingdezhen, the same area where the Yuan dynasty Porcelain Bureau had been located. The new manufactory produced porcelain exclusively for the court. For the next 238 years, Hongwu and his descendants sent officials from the capital, or sometimes palace eunuchs, to oversee the imperial manufactory and manage Jingdezhen's porcelain industry. Many of the Ming emperors took a personal interest in the production of porcelain, as did their successors during the Qing dynasty (1644–1911).[27]

Hongwu's manufactory drew ceramists from the town's commercial labour force and unskilled workers from rural and urban households. In the fourteenth and fifteenth centuries, labourers at the imperial manufactory were temporary conscripts compensated with allowances of rice. The skilled artisans had three-year part-time contracts. When they were not manufacturing for the government, they could produce commercially for themselves.[28] Most of the unskilled workers were employed in firing – a hot, demanding, and unpopular job.[29] As southeast China's economy grew and became cash-based, Jingdezhen's commercial industry expanded, and some craftsmen bought themselves out of conscription duties. The imperial manufactory switched to an entirely wage-labour workforce in the sixteenth century.[30]

Throughout the Ming, some ceramists produced for the emperor in the imperial manufactory on Pearl Hill, while others made court wares at their private workshops. Production took on an organizational form that remained in place until the imperial manufactory finally closed in 1910.[31] In the late winter and early spring, officials and ceramists made design decisions and prepared pigments and glazes. Manufacture took place from the spring through to the late autumn, with the summer months the busiest time. Firing was most successful during the late summer.[32] Potters who worked for the emperor used an official clay (*guantu*) that commercial ceramists were prohibited from using, continuing a practice begun by the Yuan court.[33] Official clay came from Macang Mountain until 1583, when the site's resources were exhausted. The imperial manufactory then procured clay from Gaoling (the 'high ridge' that gave kaolin clay its English name), where china stone was mined continuously until 1969.

Chinese ceramists were adept at using a single resource for a variety of ceramic purposes. The china stone that produced porcelain bodies could be used for glazes with the addition of limestone. Producers manipulated kiln atmospheres and firing temperatures to achieve different effects, and experimented with kiln construction.[34] Over its long history, the porcelain capital used more types of kiln than any other ceramic site in China: dragon kilns, mantou kilns (typical of North China), gourd-shaped kilns, and egg-shaped kilns (the latter is sometimes called a Jingdezhen kiln).[35] By 1597, ceramists had developed a consistent strategy to manage firing. Loaders put greenware in saggars, which were stacked in rows with small gaps or fire channels between them. Inside the kiln, the first two rows of saggars were coarse wares, followed by six or seven rows of coarse porcelain, three of mixed coarse and fine, four of the best wares, and a final three or four rows of coarse porcelain. The atmosphere of the kiln, which was mostly heavy reduction but also contained neutral and oxidizing atmospheres, determined the placement of wares.[36] In the imperial kilns, columns of saggars were left empty if their location was undesirable; in the commercial kilns, ceramists used every possible space for firing.

Jingdezhen producers also used all types of muffle kilns to fire overglaze enamels. To make porcelain with this form of surface decoration, low-firing glazes are applied to the surface of glazed and fired ceramics and matured in smaller low-temperature kilns. Jingdezhen painters learned to use overglaze enamels from northern ceramists who fled south in the late fourteenth century. During the fifteenth century, enamellers added these pigments to the best wares.[37] Officials supplied dedicated technicians with carefully allocated amounts of enamels. The early overglaze painters used the pigments to colour figures outlined in underglaze blue, creating a form of surface decoration known as *doucai* or 'joined colours'. Ming *doucai* porcelain, particularly from the Chenghua period (1465–1487), is extraordinarily refined, with the reds, greens, and yellows taking on a jewel-like brilliance against the pure white Jingdezhen porcelain. Jingdezhen's famous chicken cups, already extremely valuable by the late sixteenth century, are an example of Ming Chenghua-period *doucai* (see Plate 6).

The ceramists who worked for the imperial manufactory experimented with their media to achieve the best effects. For example, pigment producers in the early fifteenth century blended cobalt rich in iron with cobalt rich in manganese to create the most desirable shade for underglaze blue decoration.[38] They produced the best red monochrome glazes and silhouette wares in China's history, reaching a standard that later ceramists could never replicate.[39] They copied and reproduced surface decoration made at earlier times in Jingdezhen's history, and glazes from other ceramic sites. Recreations of the monochrome glazes from the Song dynasty, including the 'five famous wares' (Ding, Ru, Jun, Guan, and Ge) were particularly favoured by Chinese elites. As ceramic historians Rose Kerr and Nigel Wood note, 'Once attempted, and achieved, this principle of reinterpreting Sung [Song] dynasty and Yuan dynasty classic wares became an established feature of Ching-te-chen [Jingdezhen] porcelain manufacture through succeeding centuries.'[40]

Scholars claim that as the frugal son of a peasant, the founding emperor of the Ming sought to curb courtly consumption by mandating porcelain instead of bronze vessels for ritual and daily use.[41] But while the Hongwu emperor may have succeeded in switching the court to porcelain, his imperial manufactory operated at great expense. Wares for the emperor had to be perfect, and the court often ordered special items that were difficult to make. Hongwu's descendants sent designs to the manufactory, including patterns to use as surface decoration, three-dimensional wooden forms and wax models to create in porcelain, and actual vessels from the imperial collection for copying. Some of the wares they ordered on a regular basis were so technically difficult that out of one hundred attempts to make them, fewer than one ware was successful.[42] For less complicated porcelain, the imperial manufactory deemed between one in ten and one in one hundred pots to be of sufficient quality to merit presentation to the court. Yet even with this ratio of success to failure, the imperial manufactory managed to send enormous quantities of porcelain to the emperor, completing official orders that sometimes exceeded 100,000 pieces in a single year.[43]

To fill an order of 100,000 perfect pots, the emperor's ceramists made between one and ten million vessels that were unworthy. Until 1566, all porcelain seconds were smashed and buried, creating a treasure trove for later looters and archaeologists.[44] Even porcelain with very minor flaws was destroyed. For example, the imperial ceramists of the Chenghua period (1465–1487) are known for their refined, artistic porcelain. On one occasion, Chenghua officials smashed a saucer because one foot of one dragon painted on the vessel appeared to have six claws instead of five. The Ming emperors prohibited ordinary Chinese from using imperial porcelain: violators were to be executed and their property confiscated.[45] Yet wares made for the court still got out to the broader public, as demonstrated by the repeated proclamations of this regulation.

The court's voracious desire for porcelain catalysed Jingdezhen's development as a commercial manufacturing centre and the up-skilling of the entire population. In 1402, the imperial manufactory had to hire 20 extra private kilns to fire porcelain for the court. By 1430, officials were hiring 58 extra kilns.[46]

Already in 1487 the town was known for its non-stop manufacturing. A stone stele records: 'The town is producing porcelain for the entire country, couriers are coming and going day and night, officials are arriving from everywhere, merchants doing their business incessantly, the northern route seems to be too narrow for all of this traffic.'[47] By the sixteenth century, the imperial manufactory had grown to include a central hall with three reception rooms, a studio and residence for the official Superintendent of the Imperial Kiln, 23 storehouses, 58 kilns, four temples, two wells, and four rooms for workers to rest in.[48] Between 1529 and 1559, Jingdezhen sent at least 526,169 wares to the emperor. In 1591 alone, ceramists sent the court 159,000 pieces. Officials relied on commercial producers to fill these enormous orders. A stone stele unearthed at the imperial manufactory states that from 1608 to 1637, the imperial court relied entirely on wares made by commercial ceramists.

Commercial production centred around the imperial manufactory, with many workshops located on Pearl Hill, which was high enough to escape flood waters when the Chang River overflowed (as happened regularly). Small family-run enterprises dominated private manufacturing.[49] Proprietors who ran larger workshops recruited workers from their families, lineages, and home villages (often called native place in the scholarly literature). As noted previously, in Jiangxi and other parts of south China, a man's (patrilineal) family and lineage members usually lived in the same village. Evidence of this can be seen in the names of villages around Jingdezhen (now in some cases subsumed by it): Li Village, the Xu Homestead, the Cheng Family Village, and so on. The basic areas of porcelain manufacture – potting, painting, and firing – became associated with specific family names and native places.[50] For example, many migrants from Duchang in northern Jiangxi were surnamed Feng, Yu, Jiang, or Cao. In Jingdezhen these came to be known as the 'four big surnames from Duchang' (*Duchang sidaxing*). Duchang men specialized in throwing round or open wares, such as bowls and plates. Most commercial porcelain was thrown on stick-spun wheels: nine-tenths of all wares made in Jingdezhen during the Ming and Qing dynasties (1368–1911) were wheel-thrown open vessels.[51] Migrants from Wuzhou and Fengcheng made most of Jingdezhen's closed wares (vases, urns), migrants from Wuyuan controlled the production of pigments, etcetera.

Jingdezhen's handicraft industry required a myriad of workers: potters, painters, kiln workers, china stone miners and processors, men to make knives for trimming, transporters of unfired greenware, glaze producers, mould-makers, and men to make saggars, build kilns, pack finished wares, and so on. The promise of wages drew migrants from all over Jiangxi and Anhui provinces.

Sometimes clashes erupted between men from different native places.[52] For example, in the spring of 1541 the Chang River flooded and inundated many potteries, forcing proprietors to lay off workers. Unemployed ceramists from Leping accused the Fouliang County shopkeepers of hoarding food for local families. Riots ensued, and many were killed. Government officials described local relations this way:

> Men from Leping County in Jiangxi are drafted to work in Fouliang [County, where Jingdezhen is located]. In years of food shortages men in Fouliang, who bear the larger share of costs, neglect them or drive them out. The Leping men then sack property and rob people. Criminals from both counties gather in gangs of over 1,000 men, brawling and killing one another.[53]

To minimize conflicts between men of different native places, regulate entry into the workforce, the dissemination of technical knowledge, and relations between manufacturing sectors, ceramists established guilds.[54] The guilds standardized the sizes and types of wares that potters produced, workers' salaries and tasks, the length of the work day and work year, recruitment into the professions, and prices. They oversaw collective worship of the gods, organizing feast days with ritual parades, operas, and banquets.[55] Guilds protected technical knowledge and governed relations between industry participants. The guilds required each enterprise proprietor to maintain long-term patron–client relations with his suppliers and specialist workers. For example, the owner of a potting workshop had to patronize a specific clay supplier, knife maker (who made his trimming knives), greenware carrier, and kiln.[56] Kiln owners had to serve specific potting workshops, consistently purchase saggars from the same producer, and hire kiln loaders from the same enterprise. Subsequent owners of the businesses in question inherited these patron–client relationships. In a town densely populated with migrants and men from different kin groups, the guilds ensured the smooth functioning of production and consistency of wares. Jingdezhen's guilds regulated porcelain manufacturing from the early Ming dynasty through to 1952, when the Chinese Communist Party reorganized porcelain manufacturing and campaigned for their eradication.[57]

As Jingdezhen's porcelain industry grew, merchant syndicates from Guangdong, Fujian, and other locations became funders for commercial workshops and kilns, and established brokerages in Jingdezhen. Jingdezhen became the 'porcelain capital', and the town's name a 'famous brand'.[58] Chinese elites bought luxury ceramics from Jingdezhen, while ordinary people bought its rough porcelain. Jingdezhen was the biggest supplier of porcelain to Southeast

Asia during the Ming dynasty. Consumers in Japan and Korea also bought large quantities of Jingdezhen wares. To meet the demands of far-flung consumers, ceramists produced an incredibly wide range of finished pots. For example, Jingdezhen potters manufactured European and Middle Eastern forms, and painters copied iconography and writing in foreign languages (see Plate 7).

Europeans first encountered Jingdezhen porcelain in the late fifteenth century. From the sixteenth through to the early eighteenth centuries, they succumbed to 'porcelain disease', purchasing massive amounts of Jingdezhen ceramics.[59] Jingdezhen porcelain was such a valuable commodity that European governments sent spies to discover how it was made. Entrepreneurs across Europe tried to replicate porcelain using media such as glass, crystal, soapstone, barium, and animal bones. In 1708, the alchemist Johann Böttger finally succeeded in creating 'white gold' in Germany.[60] After this discovery, Europeans founded numerous competitor industries, and European consumption of Jingdezhen porcelain gradually declined. The American market remained strong into the mid-nineteenth century.

The tastes of China's Ming dynasty scholar-gentry class were quite different from foreigners: Chinese elites wanted antiquities. Jingdezhen ceramists replicated historic porcelain to feed this desire, and some counterfeited antiques. A man named Zhou Danquan, active during the Longqing and Wanli reigns (1567–1620), is a known Jingdezhen counterfeiter, as well as a potter, porcelain restorer, dealer, painter, Daoist, and later Buddhist monk.[61] Zhou counterfeited a Song-dynasty incense burner from the northern Ding kilns so perfectly that a major connoisseur of the day was unable to differentiate it from an original. The man learned the pot was a fake only because Zhou told him so. This connoisseur bought Zhou's incense burner in appreciation for the skill required to produce the piece. His son later sold it as a genuine antique.

Jingdezhen's success had far-reaching ramifications. The massive scale of porcelain production caused significant environmental degradation: deforestation, huge quantities of industrial garbage (kiln wasters, spent kiln furniture), and air pollution. The constantly operating kilns made the town prone to fires. Jingdezhen became a 'hub for straw sandals' (草鞋码头), densely settled with skilled and unskilled labourers. Even women, the elderly, and the disabled could find some form of work in the family-run enterprises that were responsible for most of the town's manufacture. Relatively few of Jingdezhen's residents pursued an education or activities unrelated to ceramics. However, most of the people who made significant money from porcelain were merchants from other parts of China. Jingdezhen's porcelain guilds did not concern themselves with sales; they focused on keeping each branch of the industry supplied with workers and ensuring that wares were produced properly.[62] This minimized conflicts and maintained high levels of employment while Jingdezhen had a monopoly commodity, but when European and Japanese competitors began mass-producing porcelain using machines, guild conservatism hampered innovation and kept ceramists from adapting to the new economic environment.

As the Ming dynasty failed and China fell into civil war, Jingdezhen saw fighting and destruction. Famines in 1636 and 1647 increased local residents' suffering.[63] Manchu rulers founded the Qing dynasty in 1644, and the new government reorganized the remnants of the imperial manufactory a decade later. They placed their first order, for large dragon jars over three feet in diameter, as well as other items less difficult to construct, in 1654. Ceramists were unable to make the jars. They did not succeed in filling the court's orders until 1671.[64] In 1674, Ming loyalists sacked the town, razing workshops and dwellings, destroying kilns, and killing locals. Two-thirds of the ceramists fled.[65]

After the Kangxi emperor, the fourth emperor of the Qing dynasty, assumed the throne (r. 1662–1723), he took a personal interest in Jingdezhen. Kangxi appointed officials from the imperial household to manage the porcelain industry and run the imperial manufactory in 1680, initiating a century of good governance.[66] Like his Ming predecessors, the Kangxi emperor had commercial ceramists make imperial wares in addition to the artisans at the court manufactory.[67] The porcelain industry recovered rapidly and after 1682, Jingdezhen underwent striking demographic growth. Merchants and speculators bought up property in the town centre, throwing out the elderly and infirm and renting their dwellings to ceramics manufacturers.[68]

By 1712, Jingdezhen had a population of more than a million, 3,000 working commercial kilns, and 6,000 water wheels powering wooden hammers that pulverized china stone.[69] The annual quota of porcelain for the court was 17,000 pieces of first-class round wares (bowls, plates, etc.) and 7,000 second-class pieces.[70] Pere d'Entrecolles, a missionary and industrial spy for the French government, described the city and its residents:

> There are 18,000 families in Jingdezhen. The establishments of some of the larger tradesmen occupy a vast area, and contain an enormous amount of workers. It is generally said that there are more than a million souls here, and that ten thousand loads of rice and a thousand pigs are consumed daily. Jingdezhen is about three miles of frontage along a beautiful river (the Chang), and not a heap of houses as you might imagine. The streets are drawn in straight lines, which cut each other and intersect at regular intervals. But all the space there is occupied; the houses

> are densely packed and the streets are too narrow; in travelling them one seems to be in the middle of a carnival. One hears on all sides the cries of the porters trying to make a passage. One sees also a large number of temples of idols which have been built with a great deal of expense ... Costs are much higher at Jingdezhen than at Raozhou, because everything that is consumed there has to come from somewhere else, even to the wood necessary for firing the kilns. But not withstanding the expense of living, Jingdezhen is the home of a mass of poor families who wouldn't be able to subsist in the surrounding cities. One also finds here many young workers and weaker people. It is the same way for the blind and the cripples who spend their lives grinding pigments ... One sometimes sees in this large port up to three rows of boats, one behind the other. Such is the spectacle that presents itself to view when one enters Jingdezhen through one of the gorges. The whirling flames and smoke which rise in different places, make the approach remarkable for its extent, depth, and shape. During a night entrance, one thinks that the whole city is on fire, or that it is one large furnace with many vent holes.

A customs official at Jiujiang later wrote, 'during the reign of Kanghsi [Kangxi], China's Augustan age, these porcelain factories reached a stage of development both in magnitude as well as in excellence and volume of output which they have never since equaled.'[71]

Kangxi's successor was the Yongzheng emperor (r. 1723–1735), who was even more involved with Jingdezhen's porcelain manufacture. Yongzheng established an experimental laboratory at Jingdezhen which developed new glazes and overglaze enamels. The Yongzheng emperor wanted copies of Song dynasty *qingbai*, Longquan celadons, and the five famous wares, and Ming dynasty porcelain from the Yongle, Xuande, Chenghua, Jiajing, and Wanli periods. He also ordered ceramic copies of carved lacquer, burnished bronze, and hardstones. The quantity and types of porcelain reproduced skyrocketed, and antique replicas reached new heights. Tang Ying, Supervisor of the Imperial Kiln from 1728 to 1756, recorded 60 different types of replicas made during one year of the Yongzheng period, and 107 different types made in two months of the subsequent Qianlong reign.[72] Jingdezhen also had a number of commercial potteries where ceramists specialized in producing specific kinds of antique replicas. These private potters used the same raw materials that were used in the imperial manufactory, and purchased original imperial porcelain on the market to make copies.[73] Supervisor Tang protested the sale of imperial seconds because he feared that commercial ceramists would make counterfeits, but the Qianlong emperor overrode his objections, perhaps because the emperor knew that imperial seconds were already for sale in Jingdezhen.[74] The imperial manufactory continued to sell its seconds until it closed in 1910.

Labour specialization reached peak levels during the Qing. The Fouliang Gazetteer describes 23 different types of work at the imperial manufactory.[75] For example, separate potters made large bowls, saucers, dishes, vessels for alcohol, and urns. Porcelain painters were divided into those who worked in overglaze and underglaze pigments, and subdivided into those who painted outlines and those who filled in outlined forms. Painters then specialized in particular figures: dragons, women, birds, flowers, and so on. One artisan painted the banding on round wares. A special calligrapher painted seals, marks, and inscriptions on the foot of vessels. The decorators who gilded porcelain formed an entirely different category of worker. Special groups of craftsmen were dedicated solely to the production of cobalt and charged with selecting the best pigment. They washed the cobalt, roasted it in a small kiln for 24 hours, ground and sieved it.[76] Affiliated tasks were equally elaborate: greenware and porcelain carriers, three types of saggar-makers, producers of bamboo sieves, porcelain wrappers, two kinds of worker who collected fuel for kilns, and so on.

The commercial industry imitated, and exceeded, the imperial manufactory's specialization.[77] In 1743, Tang Ying reported to the emperor that Jingdezhen had 100,000 independent artisans.[78] In 1832, officials reported hundreds of thousands of ceramists.[79] Private enterprises competed with the imperial manufactory in size, complexity, quality and quantity of wares. Clay was supplied by the clay guild, which monopolized distribution to potting workshops. Potters were differentiated according to whether they made open or closed wares, their method of construction, the type of wares they made, and the wares' quality. There were 17 types of round-ware potters and 11 types of closed-ware potters in the mid-Qing.[80] A typical open-ware workshop had 10–15 workers, related by kinship and native place. These included the thrower, trimmer, and glazer, known as the 'upper three feet' (*shang san jiao*); the moulder, who used a clay mould to standardize ware size after throwing; the foot carver, glazer, painter, greenware loader and porter, apprentices, and the man who managed supplies and accounts. Glazing methods included dipping, pouring, brushing, and blowing by mouth through cloth-covered bamboo tubes.[81]

Potting workshops took their wares to commercial kilns for firing, patronizing wood-burning kilns for high-quality wares or bramble kilns for coarse porcelain. A wood-burning kiln had ten distinct kinds of workers.[82] Firing and loading the kiln were two different industry branches, carried out by workers from different enterprises. So many centuries of intensive production had decimated local forests, and pine logs for the kilns were shipped from 160 kilometres away.[83] A kiln could be fired between 60 and 80 times, after which it had to be rebuilt lest it collapse.[84] Since the Yuan dynasty, the Wei family had monopolized kiln building, but during the late Qing dynasty a Duchang family named Yu also began constructing kilns. The Yus had been employed by the Weis as labourers, and then opened an independent kiln-repair business. A kiln owner named Feng, also from Duchang, asked Yu to build him a wood-burning kiln. At first Yu refused, because he was afraid of the Wei family. But Feng persisted, saying that he would handle the Weis. Yu built the kiln, and the Weis came to complain. But since the Wei family had more work than they could handle, they ultimately agreed that Yu could build kilns, thus ending their long monopoly.

During the eighteenth century, porcelain decorated with overglaze enamels became increasingly popular. The number of small workshops with enamelling kilns had soared by the end of the Qianlong period (1735–1796).[85] Overglaze enamelling workshops were differentiated by the colour palette the painter used, the kind of figuration the painter did, and the grade of porcelain produced.[86] The Yongzheng emperor (r. 1723–1735) played an important role in popularizing a new palette of overglaze enamels known as famille rose or *fencai*.[87] These colours, which included a range of pinks, were introduced to China from Europe around 1685. European artisans used the colours for cloisonné, but Chinese craftsmen adapted them to porcelain. Pigments came from tiny specialized workshops, typically run by husband-and-wife teams, most of whom came from Wuyuan.[88] They made 12 basic colours that could be combined into more than 100 shades. The pigment-maker combined small quantities of gold, silver, copper, iron, tin, lead, cobalt, potassium, sodium, quartz and other materials with a corrosive liquid that dissolves metals (aqua regia). Producers purchased materials from a local pharmacy, or from Foshan in Guangdong.[89] Workers ground pigments with mortar and pestle for one month, a low-paid form of labour that could be done by the elderly, the very young, the infirm, and the crippled.[90]

Another branch of porcelain manufacturing that grew popular during the eighteenth century was fine gold in the form of filigree, powder, and leaf.[91] The Wang family from Huizhou were the first in Jingdezhen to make fine gold for decorating. Gold quickly became an indispensible part of famille rose porcelain, and the Wangs expanded their operations to ten workshops, employing more than 140 people. Their Jingdezhen enterprise existed for seven generations. After Japanese and German gilt arrived in Jingdezhen during the nineteenth century, the Wangs' fine gold business took a hit. However, some producers preferred the Wangs' gold, because their materials stayed shiny and retained their colour over time. Members of the Wang family continued to make fine gold for ceramics until 1985, when the last man who knew how to produce it died.

Porcelain seconds sellers formed a branch of ceramics that existed as part of the commercial industry.[92] Seconds sellers, known as *zhoudian*, bought old, inferior, and flawed porcelain, mended it, and resold it. *Zhoudian* were differentiated by those whose methods were more permanent and included refiring, and those who mended pots using temporary cosmetic tactics. Techniques included grinding, cutting off defects and broken parts, covering flaws with enamels, and filling cracks with limestone or other filler. Seconds sellers were recognizable by the large bamboo baskets they carried to collect wasters throughout the city. The sector took its name from the part of town where most of the seconds shops were located, at the north on the sandy banks (*zhou*) of the Chang River.

Other kinds of work included making brushes for porcelain painters, using rabbit, chicken, and rat hairs as well as hemp; producing flux for glazes; making bamboo strips for various kinds of tools; porcelain selecting; and so on.[93] As ceramists developed new ware types, labour specializations proliferated. One type of ware that locals pioneered in the late Qing was known as 'bodiless' porcelain (*tuotai*), a ceramic potted so thinly that it was translucent.[94] Manufacturers in this new sector used a refined kaolin clay called *huashi* to make the ware. Bodiless porcelain manufacturing meant new mines, potters, decorators, and so forth.

The main place where porcelain was sold in the Qing dynasty was Porcelain Street (*ciqi jie*).[95] Porcelain Street was full of shops selling every type of ware. Locals ran some porcelain dealerships and brokerages, but most were operated by people from other parts of China. Many of Jingdezhen's native place associations had halls on Porcelain Street. Merchants from Fujian, Guangdong, and elsewhere needed establishments where they could stay during their purchasing visits, as Jingdezhen had no hotels. For example, the Huizhou native place association compound on Porcelain Street included a temple, a stage for opera performances, an altar for the spirit tablets of the association's founders, and several halls.[96] The association owned a number of storefronts, as most of its members were merchants.

Porcelain brokers were the sector of the industry most likely to become rich.[97] Kiln owners were the second segment of the population who typically had high incomes. The wealthy supplied credit to small-scale enterprise owners and ordinary ceramists, lending money at high rates of interest, and so increasing their economic dominance. Ceramists could get advances from workshop owners at the end of the work season.

Over the eighteenth century, Jingdezhen porcelain manufacturing experienced a golden age in terms of the quantity and quality of wares manufactured, the size and degree of specialization of the workforce, and the overall economic wellbeing of the populace. But in 1786, after a century of significant investment by the imperial government, the Qianlong emperor abolished the practice of sending officials from Beijing to supervise Jingdezhen's porcelain industry. He put the superintendent of customs at Jiujiang in charge, and made the Raozhou prefect and a Jingdezhen official responsible for porcelain transport.[98] This governance structure remained in place through the end of the late imperial period. The decision to remove supervision of the porcelain capital from the central to the provincial government marked a turning point.

From the time that China's emperors first took an interest in Jingdezhen's porcelain, the imperial government deepened and broadened its role in porcelain manufacturing, investing heavily in experimentation and design, production techniques, and management. Production for the court prompted the organizational arrangements that enabled Jingdezhen's commercial prowess, and the emperor's demands stimulated innovations in technology, ware types, glazes, and decorative techniques over hundreds of years. No other ceramics site, and no other town in China, received such lengthy and intensive attention from the centre as Jingdezhen. As one historian writing about Jingdezhen in 1989 put it, 'the porcelain industry's fortunes depend on the national government's policies. As the nation flourishes, porcelain flourishes, and Jingdezhen flourishes. We have seen this proven in history, and so it is today.'[99]

# 3

# Decline and Disarray, 1780–1948

In 1934, the industrialist Du Zhongyuan (1897–1943) investigated Jingdezhen's porcelain industry at the invitation of the Jiangxi Provincial Government. Du was a member of China's new capitalist class. He ran a successful mechanized porcelain factory in northeast China until the Japanese seized it during their 1931 invasion of the region. Du wrote a comprehensive report about Jingdezhen, which I quote at length here:[1]

> Jingdezhen is the single most famous porcelain production site in China, the first place to produce porcelain in the entire world. Because of its glorious history, Chinese citizens are of course very concerned about where Jingdezhen's porcelain industry is going. Here I write a report, for those who care to consider and discuss . . .
>
> In a new-style porcelain factory, clay-making, potting, saggar production, surface decoration, firing, quality control, and wrapping are done within a single firm. In Jingdezhen, separate small-scale private enterprises, mostly family-run, perform each of these tasks. Forms (bowls, vases, teapots, tiles, etc.) are made in potting workshops. Potters do not make their own clay, they buy it from clay sellers. Four families from Yaoli, named Wu, Liu, Li, and Shao, sell clay to all of Jingdezhen's potters. When these families make clay, they do not levigate out all foreign substances from the china stone. They make an inferior product, but earn high prices because of their monopoly. In fact, the four families purposely restrict the amount of clay they produce to keep prices high. They make a substantial annual income. The government should take over clay production and set up a mechanized raw materials factory, to standardize and improve the clay and sell it to private potting households.
>
> Jingdezhen potters throw by hand. Potters work in teams of three, and each team produces about ten boards of greenware per day [see Plate 8]. A board holds seventeen bowls or dishes. The potters' guild established this rate of production generations ago. Even using handicraft methods, the potters should be able to produce more. However they cite guild rules and refuse.

Potting workshops fire greenware in saggars, but they cannot make their own. They have to buy them from specific saggar producers with whom they have patron–client relations. Jingdezhen's guild rules give saggar-makers exclusive rights to produce for a particular set of kilns for one year, regardless of whether they make good or bad saggars. Saggars should be made from refractory clay that can withstand very high temperatures, otherwise they melt during firing and ruin the entire kiln. In Jingdezhen, saggar-makers regularly use fusible materials because they are cheaper than the refractory clay. The saggars melt, and the whole kiln topples. This happens frequently.

Jingdezhen has about 100 wood-firing kilns [see Plate 9]. They all belong to a few rich families from Duchang. Each potting household pays a firing fee before firing begins. The kiln owners are not responsible for the results. Most kiln owners do not live in Jingdezhen. They leave firing in the hands of kiln managers . . . The kiln managers charge the potting workshops another fee called 'meat money'. Potting households have to pay this fee lest the kiln manager tell the workers to put their greenware in a part of the kiln where it would underfire or overfire, or load the kiln so the saggars fall and ruin the load. Lately the firing business has declined and kiln owners are seeing less profit. They've begun building bigger kilns. To fire a bigger kiln so that all of the greenware becomes porcelain, the kiln master has to raise the temperature. The saggars cannot withstand the heat, so they melt, causing the columns and rows to fall. The kiln owners don't care if the potting households' loads are ruined, they've already been paid. In fact, the kiln owners are closing their kilns to raise the firing fees. All the kiln owners agree to fire only two or three times each month, instead of the usual seven or eight times. The potting households can't wait to have their greenware fired, so they pay higher fees – or go out of business because their overheads are too high. As fewer and fewer households pot, this puts pressure on the saggar-makers and kiln owners.

Internal problems are not all that's wrong with Jingdezhen's porcelain industry. Jingdezhen has an inadequate transportation system. The car I took from Nanchang got stuck on the muddy road, and an eight-hour journey took eleven hours.[2] Most porcelain is shipped out by the Chang River. In the autumn, the river is shallow, making transport very expensive [because men have to pull the boats with ropes, wading in the river or walking along the banks]. The government should build a railroad for the porcelain industry. Such a railroad could carry tea from Wuyuan and Qimen as well as Jingdezhen porcelain.

The world is facing a global depression, which affects markets inside and outside of China. Jingdezhen's future looks increasingly dark. In a town that once had 300 kilns operating, I saw 12 working kilns, out of a total of 100. The number of potting workshops has dropped from 4,000 to 1,000. The workforce has shrunk from 200,000 workers to 40,000. More and more people, from kiln owners to labourers, cannot sustain themselves by working in porcelain and are forced into banditry.

Jingdezhen's streets are filthy, full of trash, excrement, and rats. The people are diseased and hungry. There is not a single well-fed child in the whole town. Some fat people were travelling in my party. The townspeople followed them around, staring, because they'd never seen anyone so obviously rich before. Opium, prostitution, and bed bugs are everywhere. Epidemics of infectious diseases occur every year.

If the government does not open its eyes and ears, there is no way to save Jingdezhen. Since the Republic of China was founded [in 1912], local officials have been left in charge of Jingdezhen. They are incompetent or corrupt, totally incapable of improving the social order. Ceramists rigidly adhere to guild rules in the hopes of solving their problems, or they fight, including with weapons. The strong bully the weak, the masses attack the isolated, the porcelain industry is in shambles, and the place is a wreck.

Massive government intervention is Jingdezhen's only hope. The government should set up a porcelain management bureau to organize the industry and improve treatment of workers. It should build a raw materials factory to supply clay, pigments, and all necessary materials to ceramists. The government should build and operate a comprehensive, mechanized porcelain factory. Wood firings should be replaced by coal, to lower overheads and prevent kiln closures. The government should set up cooperatives for potting households, to assist them in buying materials and selling their wares. It should improve the transportation infrastructure, and invest in education for ceramists and public sanitation … Excluding the railroad, which I leave for other experts to assess, I estimate that the government can do everything for 520,000 yuan.

At the end of the last chapter, Jingdezhen's porcelain industry sustained a large, professional workforce and produced a vast range of high-quality wares. China's central government invested in ceramics manufacturing, ordered large quantities of porcelain, and administered production. According to Jiang Siqing, a scholar and observer writing in the mid-1930s, 'the imperial court cared about porcelain for the government, not about sales. The government's interest was artistic rather than commercial. The court was willing to spend to get what it wanted, including the cost of sending a central government official to oversee production.'[3] Porcelain manufacturing flourished under the imperial court's supervision, growing rapidly, professionalizing, and producing remarkably sophisticated wares completely by hand.

How did the flourishing eighteenth-century porcelain capital become the wreck that Du Zhongyuan described? A major factor was the withdrawal of central government support. From 1004 until the nineteenth century, the overall trend for the porcelain industry was increasing levels of state intervention. The only significant exception occurred at the end of the Ming dynasty, when the state lost the power to consume porcelain or administer production and the imperial manufactory closed. Yet because Jingdezhen produced a monopoly commodity for consumers around the world, the industry suffered no permanent damage from the short interregnum. The Manchu founders of the Qing dynasty reorganized the imperial manufactory, and Jingdezhen's global dominance continued for another next six decades, with central government officials supervising the town's lucrative commercial production.

As we noted in the previous chapter, in 1708 the alchemist Johann Böttger, who had been imprisoned by the Prince of Saxony until he produced gold, made a high-fired ceramic that resembled Jingdezhen porcelain. With state support, Böttger founded a competitor industry, Meissen. Many other European ceramics manufacturers were founded afterwards, and Jingdezhen lost the European china market. Even so, its industry remained strong with state investment and management, and consumers in the Americas, Asia and at home. The city retained a healthy economy until the Qing government decreased expenditure on porcelain manufacturing in the nineteenth century. At the same time, officials increased taxation on Jingdezhen wares, and permitted first the sale and then the domestic production of porcelain by foreign manufacturers.

The Qing dynasty collapsed in 1911, and a weak new government, headed by the Nationalist party (or KMT), struggled to control the country. In an effort to find the resources to govern, the new Republic of China raised taxes on Jingdezhen porcelain, exacerbating the industry's difficulties. The central government left the administration of porcelain manufacturing to provincial officials. The Jiangxi Provincial Government, hoping to reignite the province's former economic powerhouse, commissioned Du's investigation. Provincial officials did their best to follow Du's recommendations, establishing a Porcelain Management Bureau and asking Du to oversee all of Jiangxi's ceramics sites. Du implemented a number of reforms, standardizing kiln size, working to eradicate patron–client relations and other restrictive guild practices, and building a new mechanized porcelain factory with coal-burning kilns – but not in Jingdezhen. The new factory was in the port city of Jiujiang, where porcelain could be transported directly onto the Yangzi River, and managers would not have to fight workers' allegiance to customary practices.[4]

Du served as head of Jiangxi's Porcelain Management Bureau for a short time. In 1935 he was arrested and imprisoned for writing 'anti-Japanese slander' in a leftist journal. Du was not allowed to resume his post after he got out of jail. Jiujiang's mechanized porcelain factory closed after the Japanese invaded in 1937.

## Government disinvestment in Jingdezhen (1786–1948)

Du Zhongyuan dated poor state management of Jingdezhen's porcelain industry to the founding of the Republic of China, but as we know the central government began reducing support in 1786, when the Qianlong emperor (r. 1711–1799) terminated the practice of sending central government officials to manage the imperial manufactory and porcelain industry. Qianlong placed the head of customs at Jiujiang in charge, assisted by two provincial officials.[5] In 1799, the state cut in half the money it spent on Jingdezhen porcelain.[6] In 1806, the court again cut the imperial porcelain order in half, leaving the town with a huge overstock.

Cost savings partly explains why the government made these decisions. One proximate expense that the Qing faced was a millenarian uprising known as the White Lotus Rebellion (1794–1804). In addition, many historians argue that a century of peace and good management had caused China's population to expand dramatically, causing serious demographic pressures. The full significance of this increase was felt during the devastating famines and violent popular revolts that punctuated the nineteenth century.

From the late 1830s until the government fell, the Qing dynasty combated both domestic rebellions and foreign incursions, leaving little time for attention to porcelain. The wars with foreign nations began when the Daoguang emperor (r. 1820–1850) decided to ban the sale of opium in Canton in 1839. The British were the main suppliers of the drug, and they initiated the first Opium War (1839–1842) to keep their market open. A second Opium War took place from 1856 to 1860. China lost both of these wars, and was forced to open treaty ports, surrender territory, and make other concessions. Other imperialist nations quickly negotiated their own 'unequal treaties' with the Qing, including the United States (1844, 1858), France (1844, 1858, 1860), Sweden (1847), and Russia (1851, 1858, 1860). The Qing conceded extraterritoriality to foreigners, gave up the power to impose tariffs on foreign goods, and opened ports to foreign trade.

In 1850, a thrice-failed candidate for China's civil service examination started a civil war that lasted 14 years: the Taiping Rebellion (1850–1864). The Taiping armies won control over more than a third of China. They took Jingdezhen in

1855, razing the imperial manufactory and other parts of town. Thousands of ceramists were killed and many others fled.[7] Taiping troops occupied Jingdezhen for most of the period between 1855 and 1861, and fighting between them and the Qing armies continued in the area until 1864. When the Qing finally retook Jingdezhen, the town was devastated. In 1866 the central government rebuilt the imperial manufactory, but it was significantly diminished. According to an 1882 visitor, 'Today the Imperial buildings fall in ruins, the workrooms are badly kept, and the workmen, although recruited among the more skilful, receive a pay hardly more than that which is obtained in private factories.'[8] Production figures from 1870 to 1908 show that only 25 per cent of the imperial manufactory's output was high quality, with 57 per cent second rate and 18 per cent wasters.[9]

The Qing introduced a transport tax (the *lijin*) to Jiangxi Province in 1857, to raise money for its military campaign against the Taipings. All Chinese import or export items paid transport tax when they passed through customs barriers, typically at a rate of 10 per cent of the item's value. Officials taxed Jingdezhen porcelain at 12 per cent.[10] By contrast, the Qing taxes on foreign porcelain were between 2.5 per cent and 7.5 per cent of its value, depending on where the porcelain was going.[11] The result was that Jingdezhen porcelain cost more than imported porcelain in many of its domestic markets.

Several other domestic groups fought the Qing in the nineteenth century, including the northwestern Muslims (1862–1877), the Nian (1851–1868), and several Yunnan ethnic minorities, who, like the Taipings, established an independent (albeit much smaller) state (the Pacified Southern State in Dali, 1856–1873). As the imperial government combated these groups, foreign powers continued to demand concessions. In 1861, the Qing opened Jiujiang, the port from which most Jingdezhen porcelain shipped en route to domestic and international markets, as one of several new treaty ports. The state also gave foreign ships the right to steam or sail freely on the Yangzi River, excused foreign goods from the transport tax, and lowered the import duty on foreign goods.[12]

Preoccupied, China's central government neglected the country's flood control infrastructure during the late nineteenth and the early twentieth centuries.[13] North, central, and south China experienced repeated episodes of severe flooding. Floods in Jingdezhen destroyed kilns, workshops, residences, and street markets.[14] By the 1890s, years of flooding and food scarcity in the Yangzi region had left a sizeable proportion of the population completely destitute.

In 1894, the Qing lost the first Sino-Japanese war, and foreigners gained the right to operate factories in China. The Japanese and the British built mechanized porcelain factories in Dalian, Shanghai, and elsewhere.[15] Jingdezhen porcelain lost more market share. After the Qing fell, the Nationalist government changed the names of the taxes on Jingdezhen porcelain, but maintained and increased the total amount. A 1926 observer wrote that taxation:

> made German porcelain cheaper than Kiangxi [Jiangxi] porcelain in Peking, in spite of its much higher production cost. The German product is imported through Tientsin [Tianjin] and is liable to only two regular taxes, the import duty and the octroi tax on entering Peking; whereas porcelain from Kiangsi Province, in traveling a distance of only eight or nine hundred miles through interior China, may be subject to as many as sixty different imposts en route, depending in both number and amount almost solely on the whim of the war lords through whose areas the product has to pass.[16]

Several local officials tried to revive Jingdezhen's porcelain industry during the early twentieth century. In 1907, a former superintendent of the imperial manufactory organized a company to take over the site. He asked the central government for a dispensation that would lower tax on Jingdezhen porcelain to 5 per cent.[17] The state refused. In 1910, provincial officials and private businessmen formed the Jiangxi Porcelain Company, with the goal of mechanizing Jingdezhen. Officials provided only half of the promised initial investment, and the industry remained handicraft.[18] In 1916, Yuan Shikai commissioned 40,000 pieces of high-quality porcelain from Jingdezhen a few weeks after declaring himself emperor of China.[19] The impact of Yuan's order was as short-lived as his three-month reign. In 1929, China's Ministry of Education approved the founding of a professional ceramics school in Jingdezhen. Building began, but bandits attacked the site and stole its capital, and the project ended.[20] The 1930s saw two other efforts to 'save' Jingdezhen, one of which was Du Zhongyuan's. Both failed.

## Foreign porcelain (1850–1945)

Competition from foreign porcelain and ceramic suppliers exacerbated the government's reduced support for Jingdezhen. The English, Germans, Japanese, Russians, and Americans all sold porcelain in China, including in Jiangxi Province.[21] After the Qing was forced to make concessions to these foreign powers, their machine-made porcelain outsold Jingdezhen wares in some parts of China. The first three decades of the twentieth century brought the First World War, the Soviet Revolution, and a worldwide depression, all of which

**Table 1** Quantity and value of Jingdezhen exports, 1919 and 1930

| Year | Quantity | Value |
|---|---|---|
| 1919 | 410,545 *dan* | 4,533,052 *taels* |
| 1930 | 221,499 *dan* | 2,566,230 *taels* |

Jiang, 景德鎮瓷業史, page 190.

Note: One dan is 50 kilogrammes or 110.25 pounds. A tael is also a unit of weight, but when applied to silver was a unit of currency (a tael of silver was about 38 grammes). In 1919 a tael valued between $1.39 and $1.62 USD, according to the United States Department of Commerce's 1921 Commerce Report (volume 3, July–August 1921), page 807.

affected Europe and the United States more than China. Even so, sales of foreign porcelain in China increased.

Jingdezhen's handicraft artists tried to compete with machine-made porcelain by making coarser, cheaper goods.[22] Poor quality caused Jingdezhen porcelain to lose more market share. By the late nineteenth and early twentieth centuries, most of Jingdezhen's ceramic exports went to diaspora Chinese in Southeast Asia. During the first quarter of the twentieth century, Japanese porcelain overtook Jingdezhen wares in these markets. In the 11 years between 1919 and 1930, the quantity and total value of Jingdezhen export porcelain decreased almost by half (Table 1).

The Japanese, Germans, and Americans also sold materials for decorating porcelain in China, such as overglaze pigments, decals, transfers, and gilt. Foreign supplies competed with those made by local families, and put many producers out of business.[23] By the 1930s, foreign producers had a monopoly on some surface decoration materials, such as decals and enamels in 'foreign colours'.[24]

## Ceramists respond (1800–1948)

For much of their history, guilds were a positive force in Jingdezhen, stabilizing the industry and ordering local society. Guilds worked to ensure that every guild member made a living at porcelain production and the segmented production system functioned smoothly. But as Du Zhongyuan pointed out, after the central government quit managing the industry, Jingdezhen's guilds became problematic. Many observers complained that the guilds limited production, prevented local competition, restricted working practices, and raised costs. For example, Frank Lenz wrote in 1920 that if a ceramist was 'found doing too much and working beyond the time limit', members of his guild made sure that he was 'severely beaten'.[25] For a more serious infraction, guild members ostracized the guilty ceramist, refusing to work with him and effectively forcing him out of porcelain manufacturing.[26] When reformers or entrepreneurs tried to bring new technologies to Jingdezhen, such as machines or coal-burning kilns, the guilds obstructed them. Even small innovations generated resistance. In 1908, for example, someone developed a new technique for stamping decorations on porcelain surfaces. Ceramists rioted and killed people who used the new method.[27]

As the markets for Jingdezhen porcelain contracted, particular guilds acted in their short-term self-interest and damaged the industry. Du Zhongyuan described how kiln households fired infrequently, to raise the price of firing fees. Fewer firings per month put pressure on the potting workshops, which either had to pay higher prices or decline orders. Many potters went out of business. This in turn damaged the saggar-makers: fewer potters meant less greenware to fire and decreased demand for saggars. Ultimately, fewer potters was also bad for the commercial kilns, as it meant less need for kiln space. Another example was the greenware porters guild. Greenware carriers transported unfired porcelain from potting workshops to kilns. They prevented passers-by from bumping into and ruining the unfired porcelain by shouting warnings. If a person failed to attend to a warning and ran into a greenware porter, guild rules required that the person who inflicted the damage compensate the potter whose greenware was destroyed for the cost of *finished* porcelain, and take the porter for a meal that included meat.[28] Several twentieth-century witnesses reported greenware carriers intentionally running into passers-by, forcing them to compensate the potters and carriers.[29] The consequences included fewer wares for firing, less need for saggars, fewer finished pots for decorating, and diminished trust.

Kinship and native place alliances made guilds relations more tense. People who belonged to different kinship groups and villages had always viewed one another with low levels of suspicion. The shrinking economy and fraying production system strengthened this distrust. Fighting between people from different native places, who tended to belong to different industry segments, broke out repeatedly, sometimes on a large scale.[30] For example, in 1926 people from Leping and Duchang fought for several months, causing many deaths. Other residents hid in their houses, staying off the streets to avoid the violence. Another armed conflict erupted between Duchang and Wuzhou people, stopping porcelain production for two to three months.

In the late nineteen-teens or early 1920s, a group of kiln and pottery owners formed the Three Kilns and Nine Associations (*Sanyao Jiuhui*).[31] The association created a private, armed security force that they used to intimidate porcelain producers. The group bought off officials who objected to their tactics, and their security were allowed to wear police badges. Active through the 1940s, the Three Kilns and Nine Associations prohibited firing for two to three months at a stretch to raise firing fees and lower the cost of fuel. They refused to sell porcelain to brokers in order to force prices up. Their strong-arm tactics damaged the livelihoods of potters, fuel suppliers, kiln workers, saggar-makers, decorators, trimming tool producers, porcelain wrappers, carriers, brokers, buyers, transporters, and dealers outside Jingdezhen.

Squeezed by their bosses, guilds, and fellow residents, ceramists rioted, struck, and fought. Table 2 shows the strikes, riots, boycotts, protests, and armed conflicts in Jingdezhen between 1800 and 1948. The incident listed in 1812 was the first labour problem that Jingdezhen had seen since 1599.[32] The violence and protests in 1812, 1820, and 1823 (a struggle between owners and workers, a wrappers' strike, and guild boycotts) suggest a correlation between the government's initial retrenchment at the beginning of the nineteenth century and labour unrest. Even more striking is the number of incidents that occurred between 1900 and 1948, when the state had withdrawn all support from Jingdezhen and promoted a high rate of taxation. In this 50-year period, 31 conflicts were recorded: 25 strikes, two riots, one protest, two armed conflicts, and guild boycotts. That is an increase of 20-fold – and contemporary observers, a historian, and an ethnographer working with elderly informants in the 1990s believe that there were even more strikes than the written record shows.[33]

Two-thirds of the incidents in Table 2 were about compensation. The others were about innovations in production that threatened the livelihood of particular industry sectors, and guilds that were disadvantaged by the regulations and practices of other guilds. For example, seconds sellers bought broken or damaged porcelain from potting workshops or decorators, repaired it, and resold it in their own stores. In 1910, a fight broke out between seconds sellers and ceramic painters. The seconds sellers had started using stamps and decals instead of painting to repair the decoration on damaged porcelain. Porcelain painters protested, fighting ensued, and parties from both sides died in the violence. The conflict forced the creation of a new guild rule allowing seconds sellers to use stamps and decals instead of brushwork for surface design, but prohibiting other branches of the industry from doing so.

Only one conflict between 1800 and 1948 was not about either compensation or innovation: a 20-day strike in 1900, caused by a rumour that local Catholics were burying children alive.[34] This strike, during which ceramists captured the

**Table 2** Strikes, riots, protests, and armed conflicts in Jingdezhen, 1800–1948[35]

| Year | Incident |
|---|---|
| 1812 | Violent struggle between porcelain workers and owners, 2000+ participants |
| 1820 | Wrappers strike |
| 1823 | Guild boycotts |
| 1900 | 20-day strike, ceramists burn down Catholic church and pharmacy |
| 1901 | Two potters strikes: (a) open- and closed-ware potters; (b) coarse-ware potters |
| 1902 | Strike protesting inadequate compensation (low-quality rice) |
| 1904 | Series of strikes |
| 1908 | Riot over new stamping methods, killings |
| 1910 | Riot over porcelain seconds sellers stamping porcelain rather than painting |
| 1920 | Guild boycotts, worker protest over payment in copper rather than silver |
| 1923 | Wrappers strike |
| 1924 | Two strikes: (a) kiln workers over delayed opening of kilns; (b) low wages |
| 1926 | Two long, armed conflicts between industry sectors, enamellers' strike for better wages |
| 1927 | CCP organizes labour unions, participates in ongoing strikes |
| 1928 | Nine strikes for better wages |
| 1929 | CCP organizes large strike, attacks on hoarding shopkeepers, tax offices, tax boat |
| 1932 | Strike over government plans to establish new anti-bandit tax |
| 1938 | Strike for wage increases |
| 1946 | Two strikes: (a) wrappers strike; (b) kiln workers, potters strike for wage increase |
| 1948 | All industry strike for wage increase |

local French missionary and burnt down a Western-style pharmacy, occurred during a widespread anti-foreign, anti-Christian uprising, the Boxer Rebellion, which began 1898 and was crushed by an eight-nation alliance of European and American forces in 1901.

Ceramists used multiple methods to protest.[36] One was to slow down the pace of work. Another was to refuse to work. A third was to destroy the wares they had made. Sometimes a guild would order its members to stop selling goods (e.g., trimming knives) to an industry partner. Still another form of protest was to 'topple the kilns', wrecking unfired greenware and saggars. Workers also participated in formal, large-scale strikes.

Between 1928 and 1933, a branch of the Chinese Communist Party (CCP) helped organize strikes and protests.[37] Shao Shiping, a leader from the Northeast Jiangxi Soviet Base, first came to Jingdezhen in 1927 to organize an underground armed organization. Its members were discovered and killed by the Nationalists. In 1928, a secret CCP organization and youth league were formed, and members organized a strike of overglaze enamellers for better wages. For more than 20 days, 1,600 porcelain painters struck, and successfully obtained their demands. Fang Zhimin, a major CCP leader, arrived in Jingdezhen in 1929 and organized an all-industry strike, attacks on salt and rice merchants who were hoarding goods, and the destruction of a tax boat and office. The 10th Red Army came to Jingdezhen three times in 1930, and distributed money, food, and cotton goods to ceramists and townspeople. One final strike occurred under CCP leadership in 1932. In 1933, the Nationalist government destroyed Jingdezhen's Communist organization, executing six local leaders. Ceramists' strikes – which predated the CCP's arrival by more than two decades – continued for another 15 years.

As Jingdezhen's porcelain industry fell apart, living conditions worsened. After his 1923 visit, Harry Franck wrote this description of Jingdezhen:

> No wonder our Shanghai consulate requires a certificate of disinfection on all Kingtechen [Jingdezhen] ware destined for the United States; the whole town was as utterly crapulous as any I had seen in China; no one, nothing, could really be called clean. The stench of human excrement, of never-washed people living in sty-like dens, the mangy scalps and ulcerated skins, and all the other filth-diseases with which China, particularly its southern half, swarms, were everywhere ... [The locals] lived in hot dens succeeding one another endlessly along narrow streets, with no breathing space anywhere even by day, and by night their hovels as tightly closed in any weather as mud bricks and crude carpentering can make them.[38]

Du Zhongyuan's 1934 report shows how little had changed in the decade between Franck's and his visits. During Du's short tenure as head of the Porcelain Management Bureau, the local Merchant's Association built some roads of stone, and a small waste-water system. A resident purchased Jingdezhen's first motor vehicle in 1939.[39] But in the 1940s, the Japanese repeatedly bombed the city, causing many potters to flee.[40] By 1948, the town had fewer than 100 kilns, only eight of which were operating. The city whose kilns had once lit the night sky so brightly that the town resembled an enormous furnace was almost unrecognizable.

China during the late Qing and early twentieth century experienced troubled times. The country underwent foreign incursions, rebellions, war and two episodes of state failure. In Jingdezhen, ceramists confronted reduced state inputs, higher taxation, decreased demand, and foreign competition for porcelain markets, as well as serious bouts of warfare. Central government administration disappeared. In this world of shrinking resources, the porcelain capital's social fabric came apart. Guilds, which had enhanced the industry's functioning, became obstacles to production. They blocked technological and aesthetic innovations and prioritized their members' short-term interests over the productive long-term relationships that had been Jingdezhen's hallmark. One sector of the industry boycotted another to extort a higher price for their services, the owners of kilns and potteries formed associations to squeeze workers, and workers rioted for better pay. On the streets, violence became commonplace as men from different kin groups and native places accused one another of exploitative and immoral behaviour. As the Japanese invaded and civil war erupted, Du's call for the government to save Jingdezhen remained unanswered.

# 4

# Production and Politics, 1949–1972

When the People's Liberation Army entered Jingdezhen at 10:00 am on 29 April 1949, they met no resistance.[1] The former government had fled three days before, and the town was quiet. Bandits and pirates had disrupted commerce by land and water. The potting workshops were closed, production of china stone had halted, and kiln fuel was in short supply. Many ceramists lived in the potting workshops, which had walls on only three sides, while others stayed in rough structures built from old kiln bricks and saggars. Sewage flooded the town's streets. Chinese Communist Party administrators, or cadres, who arrived with the troops recorded that much of the local population was near starvation.

Feng Shangsong was 19 years old when the army arrived.[2] He had already been working as a porcelain thrower for eight years, four as an apprentice, and four as a contract labourer. His father had died when Feng Shangsong was four, and his mother had remarried the following year. Although 'virtuous widows' who never remarried were much praised during Feng's childhood, only wealthy families could afford to maintain a young widow for the rest of her life. Feng's mother moved in with her new husband and his family, who would not have been willing to support a boy who belonged to another descent line. Feng stayed with his paternal grandparents. Only a family in truly desperate straits would ever allow a son or grandson to be adopted into another kin group.

Feng's grandfather was a kiln master (*bazhuang*) at a commercial kiln. Firing was the make-it-or-break-it stage in porcelain production, which is why most ceramists worshipped the God of Wind and Fire. A kiln master 'watched the fire' (*kan huo*), judging when the kiln had reached the proper temperature based on the colour of the flames (kilns had no temperature gauges), determining when to add wood and whether to add wet or dry logs, and deciding the length of time required for perfect firing. Kiln masters stood at the pinnacle of Jingdezhen's labour hierarchy, and in good times were men of means. Jingdezhen in the 1930s was experiencing bad times, but working kiln masters like Feng's grandfather still had better incomes than other ceramists.

Old Feng wanted his grandson to be educated, so he sent Feng Shangsong to school at the age of seven. Jingdezhen's schools in the late 1930s and 1940s were small private ventures, based on Confucian texts and run by literate men. Feng studied the classics for three years, and then his grandfather died. Feng's grandmother took in sewing and laundry to support herself and her grandson, but could not afford to keep him in school. She found an apprenticeship for her ten-year-old grandson. Feng was too young and small to load saggars, which was the first step toward becoming a kiln master, but a boy of his age could learn throwing (*la pi* or *zuo pi* in local dialect). Young boys made good throwers because their hands were 'soft'. While a potter was not as well remunerated as a kiln master, throwing was one of three key positions in a ceramics workshop: 'first throwing, second trimming, third glazing' according to a local aphorism (*yi zuo er li san shangyou*).

Feng's youth was a time of poverty and hardship. He was too poor to own shoes. Potting ceased during the coldest three months of winter, when the clay was too hard to work with, so Feng had to find other sources of food during those times. He chopped firewood, worked as a shepherd in the countryside, and peddled bits of porcelain for rice. Feng remembered 1945 as a particularly hard year. First he fell off the roof of a kiln shed (kilns were built inside sheds, where the wood for firing was stored). Then, while he was hauling water from a well, he fell in. Later, while working in the countryside, he fell from a second-storey loft, injuring his foot so he had to limp all the way home to Jingdezhen.

For Feng, the arrival of the People's Liberation Army ushered in hope for a better life. 'The People's Liberation Army came, and the Chinese Communist Party came,' he recalled. 'To the Party, we poor people, those of us who were the children of poor people, we were just as good as anyone else.' This attitude affected Feng profoundly. He decided to 'join the revolution'.

To many Europeans and Americans, China's Communist government under Mao Zedong is associated with iconoclasm, anti-capitalist struggle, and large-scale campaigns that caused death and damage. Yet many Jingdezhen residents like Feng, who were young adults during the 1950s and 1960s, described Jingdezhen's high socialist period as a time of nation-building and government investment, which laid the foundation for the porcelain capital's development as a modern ceramics industry. Older adults spoke about new infrastructure and city expansion, high levels of employment, improved standards of living, and population growth. Those who became porcelain workers in state and collective enterprises had high social status. Many were nostalgic for the relatively egalitarian society that the CCP created.

After decades of neglect, China's new leaders reinitiated state management of Jingdezhen's porcelain manufacturing, reorganizing the industry around production and politics rather than state requisitions or sales. Officials changed the industry's workflow, technology, and output, adding architectural tiles, electrical ceramics, and porcelain for sanitary use to the decorative and daily use porcelain for which the town had been known. They reformed labour, including recruitment, apprenticeship, compensation, working conditions, and the length of the workday and work week. Party administrators regulated many aspects of residents' lives, including where they lived, what they ate, how they got married, and which pattern decorated their porcelain rice bowls. At times, ideology and activism took precedence over ceramics production. Overall, the return of big government was good for Jingdezhen and its ceramists.

## Economic recovery, 1949–1952

China's Communist Party had radical political goals: the Party hoped to build a modern, egalitarian society in which all citizens' needs were met. China's new leaders wanted to topple the wealthy elites, large landowners, and big capitalists who had benefitted most from the pre-existing social system, incorporate women into the paid workforce, and build an economy based on rational planning rather than markets. To achieve their goals, they had to convince people that familiar social structures, including the family, native place associations, guilds, and religious organizations, should be transformed. But first, the CCP had to prove that their regime was permanent. This was difficult in places like Jingdezhen, where locals had watched Nationalist government officials execute Party members only 15 years earlier.

The local Communist Party and People's Government, set up by the Army, acted quickly to get porcelain production started. Within days, officials placed an order for porcelain and initiated an intensive propaganda campaign, aimed at the owners of workshops and kilns, proclaiming the government's intention to protect business and commercial relations.[3] Cadres called a meeting for all enterprise owners and ceramists, and asked the porcelain workers to reduce their pay as a way to stimulate economic activity. Potters who worked with round wares (bowls, plates, cups) were told to take a 20 per cent pay cut, and those who worked with closed wares (largely decorative ceramics) a 40 per cent pay cut.[4] Ceramists who remembered the CCP-organized strikes for higher wages during the early 1930s must have been shocked. According to official documents, porcelain workers were barely paid enough rice to feed themselves.[5]

Jingdezhen had seven traditional banks, founded by (former) officials, merchants, and wealthy citizens. On 20 June 1949, Communist officials took them over and established a branch of the Bank of China.[6] The government prohibited the use of old currency or precious metals in favour of the new People's currency. The new bank gave low-interest loans to ceramics businesses, focusing on small and medium size enterprises. As families accepted credit and began to reopen their businesses, officials began to count the number of working kilns, potting workshops, and ceramists. During the following months, cadres repeatedly surveyed these measures of economic activity. As manufacturing recovered, officials added the number of wares produced. Preoccupation with these indices of economic health continued into the early 1990s.

The government turned to the People's Liberation Army to make the area safe for commerce. Over the next several months, the army rooted out, captured, and killed armed bandits and pirates in the area, confiscating 400 guns.[7] Soldiers accompanied merchants transporting porcelain out of Jingdezhen. Meanwhile, party administrators liaised with potential customers in large cities, persuading them to make purchases and invest in production. The city government placed several orders itself.

To ensure steady supplies of china stone and pine logs, officials adopted the solution proposed by Du Zhongyuan in the early 1930s: they put the supply of raw materials under state management.[8] This ended the china stone monopoly held by four powerful families, striking a blow to the 'feudal' family and lineage system. Officials created mining cooperatives that shared tools, labour, and transport, but kept profit and ownership in private hands. To resolve fuel shortages, the provincial government ordered farmers from the whole of Jiangxi to send pine logs to Jingdezhen. For 270 kilometres around the town, use of pine for anything other than kiln fuel was strictly prohibited. Cadres hired women to chop wood in areas with labour shortages, which was another attack on the 'feudal' family system.

In August 1949, the city government founded an experimental socialist workshop to make polychrome glazed porcelain in the old imperial manufactory. Cadres intended the new enterprise to be a model for the rest of the industry, and it became Jingdezhen's first state-owned porcelain factory, reopening as the Build the Nation Porcelain Factory in April 1950.[9] The enterprise made several formerly private workshops, kilns and parts of the old imperial manufactory into a single operation. While Build the Nation still used handicraft production technologies, had dispersed work sites, and paid workers in rice, new rules governed manufacturing. Ceramists had to guarantee that they would make a specified daily quantity and quality. Those who failed were docked pay. Those who exceeded their requirements got bonuses. The enterprise had a labour union, a barbershop, and a clinic. Workers received one rest day and one free haircut per month. In the evenings, the factory hosted literacy and maths classes for workers, six days a week, two hours per session. Build the Nation was not simply a place of employment: city officials had created Jingdezhen's first 'work unit' (*danwei*) by combining work, social services, and governance. With 822 employees, it was by far the biggest production site in town.

The advantages received by employees of the new socialist experiment did not go unnoticed. Not long after the enterprise opened, Jingdezhen's round-ware potters argued that they should return to their original wages, citing increasing sales in the wake of the autumn harvest. In October, ceramists who made coarse wares struck to protest their low wages.[10] One of the Party's responses was to organize Jingdezhen's first public criticism or 'struggle session' on 29 October 1949.[11] Locals were encouraged to speak out against 'the most hated exploiters', humiliating them for all to see. Cadres also began to register local capitalists, including small-scale owner-operators, a process that was completed in 1953.

Over the new government's first year, officials called several meetings for owners and ceramists at which they encouraged cuts to traditional payments and pushed workplace reforms. In December 1949, cadres brokered an agreement between owners and workers to cut ceramists' traditional year-end bonus by 20 per cent. Potters who made coarse wares agreed to reduce production and the length of their workday, taking in turn a pay cut. The city government mandated that bosses could no longer curse or beat their employees.[12] Over the following 12 months, round-ware potters returned to their customary wage, and closed-ware potters, whose business recovered more slowly, went to 90 per cent of their former pay.

Feng Shangsong was still working as a thrower, but because he could read and write, Jingdezhen officials gave him responsibilities working for the labour union. The government and Party were desperately in need of literate people to assist with governance. They allowed local elites to work for the new administration, so long as they did not obstruct the Party's goals.

Officials created the Porcelain Industry Company (*Ciye Gongsi*) early in 1950.[13] The Porcelain Industry Company purchased raw materials and sold them to private producers, bought finished porcelain and sold it to brokers, and managed government orders. The new administrative unit classified wares by grade and established price standards. In time, the Porcelain Industry Company became the single most important element of the state apparatus in porcelain, managing all aspects of production and distribution for what became a massive industry.

Porcelain prices dropped in the spring of 1950, to the dismay of officials and local ceramists. Forty families who had been operating ceramics workshops closed them and fled the city.[14] Apparently they feared reprisals for closing. Jingdezhen officials initiated a campaign for private producers to form cooperatives between June and December of 1950.[15] Producers would work together to procure supplies and sell wares, but revenues and ownership would remain private. Five groups created cooperatives: spoon-makers, decorative ware makers, round-ware producers, porcelain wrappers, and transporters. In September 1950, the government persuaded a firing cooperative to form. The cooperatives met monthly to plan operations. Officials approved which enterprises could join the cooperatives, and set standards for contracts, master–apprentice relations, and 'women workers issues', meaning toilet facilities. Cadres offered literacy classes to cooperative members. Yet the city lacked adequate administrative capacity to ensure that the cooperatives functioned smoothly. Four broke up after only a few months.

On 1 May 1950, China's central government passed a new Marriage Law that gave women legal equality with men. Jingdezhen cadres began to criticize the long-standing practice of arranged marriages, where parents determined who their children should marry, and promoted love marriages, in which couples themselves made the decision to marry. Officials required all couples who wanted to marry to register with the government.[16] At the same time, party members encouraged residents to adopt plain, gender-neutral garb, Mao suits (or Sun Yat-sen suits) of blue and grey (see Plate 10).

Central government officials passed the Land Reform Law in June 1950, legally abolishing landlordism and removing property from those who did not till the soil (including temples, monasteries, and other institutions). In Jingdezhen, officials prepared the population for the redistribution of land by confiscating property from three 'hated exploiters' in October 1950. Mu Shaosu had been the head of the Nine Kilns and Three Associations, the coercive owners' group described in the previous chapter. Officials took one kiln and three potteries from Mu. Lai Zhunqing had been the head of Jingdezhen's Merchants Association. From him, the government confiscated two comprehensive potting workshops (potteries with kilns). Xiang Dechai had been a senator under the Republic of China. From Xiang, officials appropriated one kiln and eight potteries. All of this property became part of the Build the Nation Porcelain Factory.[17]

Officials began land reform in the countryside around Jingdezhen on 5 January 1951.[18] Many local farmers produced 'little white bricks' (*baidunzi*) of china stone, the key ingredient of porcelain clay, in addition to growing rice. The Party sent 120 cadres to the surrounding villages, to oversee the confiscation of land, tools, furniture, houses, gold, silver, food, and grain from landlords, and their redistribution to farmers who had little or no property, as well as a modest portion to the former landlords themselves.

In February and March of 1951, the government ordered ceramics from 400 workshops throughout Jingdezhen.[19] Officials then initiated a second campaign to promote cooperatives, this time creating an office to provide support.[20] Cadres encouraged 43 households that made large-sized saggars to form a Large Saggar Cooperative. They helped 11 comprehensive potting workshops merge to become the Build China (*Jian Zhong*) Cooperative. All together, two other potters cooperatives, seven porcelain distribution cooperatives, six food producers, a glaze-maker cooperative, a group of overglaze enamel producers, and two more cooperatives of saggar-makers formed. Cadres publicly criticized guilds and native place associations as part of the campaign.[21] Official changes to traditional payments, the length of the workday, hiring practices, and the operations of the Build the Nation Porcelain Factory had already diminished the power of these traditional institutions. Cadres now terminated the practice of payment feasts, previously an important guild event.

On 21 March 1951, 30 April 1951, and June 1952, officials executed four 'evil officials and bosses' (Xiong Dexiu, Xiong Shuigen, Hu Guiyi, and Zhang Buben), a group of 'counter-revolutionaries', anti-Party troops, some heads of native place associations, and four 'non-cooperating' capitalists.[22] Official documents state that the public executions 'improved morale'. In total, city officials executed about 0.05 per cent of the city's 192,700 residents.[23] This was a much lower proportion of executions than were carried out in the nation as a whole: between 1949 and 1952, China's government executed 4 per cent of the nation's population as counter-revolutionaries and exploiters, roughly 2 million executions from a populace of 500 million.[24] In Jingdezhen, if owner-operators and other capitalists cooperated, the Party's policy was to 'use them, limit them, reform them'.[25]

Locals recalled that during the first years of the Party's tenure, many people from other parts of Jiangxi migrated to Jingdezhen. Some had been given official identity classifications such as landlord, capitalist, or rich peasant, all known as 'bad class labels', in other places. Jingdezhen officials assigned a number of these migrants better class labels, such as middle peasant or 'city resident' (*jumin*).

Others were allowed to 'draw a line' between themselves and their class-enemy families.

On 3 August 1951, city officials sent 158 cadres to make 'democratic revolution' in all the pottery workshops and china stone mines, including the state-owned Build the Nation Porcelain Factory.[26] At Build the Nation, cadres organized a 'speak bitterness' session, encouraging workers to describe their past sufferings in public. Ceramists at the enterprise spoke out against 12 employees, some of whom had informed against Communist Party members in the 1930s. Officials fired these individuals and arranged for workers to elect new management. Across all of Jingdezhen's workshops and mines, a total of 144 people were declared tax evaders, embezzlers, and thieves.

The 'democratic revolution' was only a first step to foster among ceramists a sense of shared responsibility for the industry. In December 1951, officials hosted a competition where they challenged porcelain workers to find new 'solutions' to increase production and make manufacturing more efficient.[27] Ceramists responded with scepticism. 'Small solutions are useless, and big ones are hard to find,' grumbled some. 'What solution is there for a trimming knife?' argued others.

In 1952, party administrators pushed for small private workshops to merge. The potting workshop where Feng was a thrower joined with another workshop to become the Jingdezhen Benefit the Masses Porcelain Workshop (*Jingdezhenshi qunyi cichang*). Feng became a full-time administrator, heading its new labour union. He and other officials made a strong push for all ceramists to become union members.[28] Feng also got married to a young woman from a large, poor family. After the wedding, Feng used his government connections to secure a job for his wife at the Build the Nation Porcelain Factory. He encouraged her to participate in the factory's evening literacy classes, hoping that she would learn to read and become a nurse. Mrs Feng managed to learn 2,000 characters, a remarkable achievement for a 21-year-old young woman. To work in medicine, she would need more than double that number.

Cadres tried to generate team spirit among potters by organizing events that linked porcelain production to patriotism and the War to Resist the United States and Assist Korea (the Korean War).[29] They held a second competition to 'find solutions, learn solutions, disseminate solutions', this time with better results: 13,382 participating ceramists, 222 new 'solutions'.[30] City officials set up a new workshop to develop pigments, transfers, decals, and gilt for surface decoration. By the end of 1952, the workshop produced gilt 'to an international standard'. Cadres touted this achievement widely, proclaiming the end of decades of dependence on foreign-made materials.[31]

At the end of 1952, the Chinese Communist Party declared that China's economic recovery period was concluded. Jingdezhen's economic situation had improved markedly, and porcelain manufacturing bore a strong resemblance to how it had looked in the past. Ninety-nine per cent of the economy was private porcelain businesses, and private production as a percentage of economic activity had grown between 1949 and 1952.[32] The Build the Nation Porcelain Factory and Porcelain Industry Company were Jingdezhen's only state-run enterprises. Most of the 2,000 potteries scattered throughout the city were independent concerns employing fewer than ten workers who used handicraft methods to make wares.[33] The porcelain workforce had increased to 18,262 ceramists, all of whom were still paid in rice.[34]

## From capitalism to socialism

Between 1953 and 1957, retroactively designated as China's first Five Year Plan, central government officials led China out of capitalism into a socialist planned economy. In Jingdezhen, officials assumed direction of the porcelain industry, reorganizing and expanding manufacturing. They created new wares, developed new technologies, and promoted a new aesthetics. Cadres increased efforts to bring women into the workforce, and thoroughly reformed potters' compensation. They built infrastructure, housing, and new factories, and created Jingdezhen's first public park. Officials also formed neighbourhood associations, which were a grassroots level of governance that promoted compliance with official policies. Looking back after more than five decades, older residents described the 1950s as an energetic time when 'the nation' invested in and improved their lives, and they in turn devoted themselves to the national good.

Early in 1953, cadres began holding political meetings and study sessions where they discussed the problems that capitalists caused.[35] They explained that capitalists kept porcelain manufacturing backwards, chaotic, and oriented toward profit. Capitalists had bad management practices: if one owner made money selling coarse porcelain, then all the capitalists rushed to make the same wares, regardless of what was needed. Capitalists prevented the cooperatives from increasing production and caused prices to rise. Officials mounted dance and choral performances, showed movies, and painted posters to demonstrate the benefits of socialist reform.

Government control of the porcelain industry, which had begun with state management of raw materials, expanded until private enterprise was eradicated in January 1956.[36] In 1953 the Porcelain Industry Company took over most sales, organizing 169 private enterprises into a nationalized quota system. Each workshop received a production target measured by weight, with a specified percentage of coarse and fine porcelain and goals for wastage and costs. Officials inspected their production facilities, and owners signed a contract for distribution by the Porcelain Industry Company. The government lifted old taxes from these producers. By August 1954, 95 per cent of ceramics workshops had been integrated into this state–private system. In January of 1956, officials reorganized all porcelain producers into 52 handicraft cooperatives. On 1 February, officials reorganized the cooperatives to create ten large state-owned enterprises employing 11,361 workers, and 20 collective factories employing 11,010 workers. Under the new system, the government told the reorganized 'factories' what to produce, standardizing and simplifying the traditional ware types, and adding new products such as porcelain tiles, industrial and electrical ceramics, and sanitary porcelain.[37] Similar processes took place in the surrounding countryside, where private and cooperative farms were transformed into communes and collectives by 1957. As part of communal labour, every county in Jiangxi, and some in Anhui, was required to provide china stone and wood to Jingdezhen.

Wage reform was a major component of the transition to socialism. The government set up a Labour Bureau and charged Feng Shangsong to head a work team to investigate and rationalize wages. After many months of research in all sectors of porcelain manufacturing, Feng and his team established salary grades so that labourers were paid according to their work and their technical skills. They implemented an incentive programme, awarding individuals and work teams cash prizes for productivity, and recognizing them as 'model workers' and 'model teams'.[38] Whereas under the guild system each guild member had received the same wages – which meant that all throwers earned the same amount, as did all trimmers, all glazers, and so on – now officials linked pay to work.

Cadres also wanted women to join the paid workforce. Officials walked the city's neighbourhoods, calling on women to liberate themselves by working in porcelain factories. They ordered china stone mines and porcelain workshops to construct women's toilets and establish day-care facilities.[39] Historically, men had performed the vast majority of labour, skilled or unskilled, in porcelain manufacturing. Many ceramists thought that women brought bad luck to ceramic production. Women were forbidden to come near kilns for fear their presence would cause a misfire or make the saggars topple.[40] Women's roles were traditionally small and largely informal: a wife joined her husband in cooking pigments for surface decorations in a small-scale family enterprise; women sewed the special thick gloves and vests worn by the men who opened the kilns after firing and unloaded the hot ceramics.[41] The women-to-work campaign brought 2,041 women jobs in cooperative porcelain enterprises by the end of 1955, out of a total of 10,082 cooperative workers.[42]

Mrs Jiang had married into the Jiang family in 1941 at the age of 11. The Jiangs were well off by local standards, even though Mrs Jiang's husband's grandfather had lost the family's ten storefronts on Porcelain Street, and part of the family's residence, while gambling at mahjong. The Jiang men were known for severe hereditary nearsightedness, which never stopped them from gambling. When Mrs Jiang married in, the family owned a porcelain workshop and hired men to produce tiny blue and white birdfeeders. As her dowry, Mrs Jiang brought a pair of gold earrings, which was all that her family could afford. Her husband's paternal grandmother (husband's father's mother) demanded that Mrs Jiang surrender them to her. Sixty-five years afterwards, the loss of her gold earrings still made Mrs Jiang cry. She became the lowliest member of the Jiang family, forced to rise earliest, work hardest, and eat the least. Much of her time was spent waiting on her mother-in-law and husband's grandmother.

When the Jiang family turned over their workshop to the government in 1953 or 1954, all of the able-bodied family members received jobs as manual labourers in porcelain manufacturing. Mrs Jiang's husband was given a job hauling clay from the area where it was made to the section where it was thrown. For a Jiang man, hauling clay was an abrupt drop in status, a punishment for having been capitalists before the founding of the People's Republic. Mrs Jiang also received a job hauling clay. To her, the job meant earning her own wages, having an independent social sphere, and getting away from the control of her mother-in-law and husband's grandmother. Mrs Jiang was proud to go to work.

Measures to distribute food and cloth more equitably accompanied the integration of women into ceramic manufacturing. Officials began rationing grain in March 1954, and added cotton and cloth rationing that September. Residents had to be registered with the government to receive ration coupons, and only those who worked were entitled to them. The quantity depended on an individual's grade of work.[43]

Numerous construction projects accompanied the economic restructuring. In 1953, the government dynamited the Chang River, making it deeper so that boats could transport porcelain out of the city more easily, and organized construction of the first bridge in the city's history.[44] In 1954, officials built Jingdezhen's first public park on the west bank of the river, newly accessible by the pontoon bridge, which included a reading room, library, teahouse, and music hall.[45] The government laid a new road to the east, 'the First East Road', and built along it housing, schools, and four factories.[46] Officials constructed more housing in the villages close to Jingdezhen, beginning these areas' transition into urban neighbourhoods. Many parts of Jingdezhen bear names that show their origins as villages: Li Village, New Village, Worker Farmer Village, and so on. In September 1955, officials started a public bus system, beginning with two buses and increasing to four by the year's end.[47] Other improvements followed: the city's first running-water provider in December 1956; 500 street lamps in 1957.

China's central government was desperate for foreign currency, and the Jiangxi Provincial Government pushed Jingdezhen officials to recover or find new international customers for porcelain. At this time, the United States, France, Germany, and Japan controlled 88 per cent of the international ceramics market.[48] Jingdezhen cadres designated seven factories to produce export ware. They liaised with potential buyers in traditional markets (Hong Kong, Singapore, Malaysia, and Macao) and new ones in the Soviet Union and Eastern Europe. They took Jingdezhen porcelain to international exhibitions in Poland, Indonesia, and other countries.[49] Thirty Soviet experts came to Jingdezhen and taught Jingdezhen ceramists mechanized slip-casting methods.[50] By 1956, Jingdezhen was exporting porcelain to Holland, Switzerland, Sweden, Egypt, Mongolia, India, Vietnam, Burma, and elsewhere.[51]

Officials continued to use 'solutions' competitions as a way to involve ceramists in changing technology and improving manufacturing. Cadres displayed the results of one at a public exhibition in October 1954: new carts for hauling china stone, foot-pedal potter's wheels, plaster-of-Paris moulds to standardize the size of round wares, trimming knives, banding wheels, vehicles for loading kiln wood, and several new production methods, including painting surface decoration with two brushes, two-handed glazing, a technique for stoking kiln fires, strategies for stacking greenware in kilns.[52] The range and quantity of new technologies indicate that officials were able to motivate significant participation. Foot-pedal wheels were immediately put into use at the Build the Nation Porcelain Factory, and officials hailed the state-owned enterprise as 'half-mechanized'. By the end of 1955, one-quarter of all the potter's wheels in Jingdezhen were foot-pedal wheels.[53]

Kiln-building became a state task during the first Five Year Plan, and city officials began experimenting with a coal-firing kiln at a branch of the Build the Nation Porcelain Factory (later designated the Universe Porcelain Factory) in 1954. The new coal kiln was tunnel-shaped, unlike the traditional egg-shaped wood-burning kilns, and wares rotated in and out on carts. After two years of effort, workers achieved a successful firing in 1956.[54] Thirteen enterprises adopted coal kilns, and 32 changed to combined wood-and-coal kilns by 1957. A group of ceramists who had been kiln managers and loaders for wood-burning kilns suddenly found their expertise worthless. But the Party averted labour unrest by insuring that these ceramists remained employed, albeit at different tasks.

In 1956, the central government ordered two chemical plants from Shanghai to move to Jingdezhen, consolidating them into the nation's first factory devoted to making ceramic pigments.[55] Central officials sent experts from the Chinese Academy of Science and the Light Industry Ministry to Jingdezhen, to research the chemical components of clay bodies and glazes, define scientific units of measurement, and determine precise firing temperatures.[56] Local officials founded a Ceramic Research Center (later renamed the Light Industry Bureau Ceramics Research Center) to host the Beijing scientists and train local experts. Feng Shangsong, who was serving as Deputy Head of the Jingdezhen City Wage Reform Office and Head of the Jiangxi Porcelain Industry Company Labour and Wages Section, travelled to Shanghai to recruit staff for the research centre. 'Jingdezhen didn't have a single college graduate at the time,' he told me. 'In fact, very few people were even high school graduates.'

On 1 January 1954, the city government opened a new Ceramics Museum. The museum showed ceramics from the Neolithic period through the Republic of China, new ceramics for the textile industry and electrical use, and an exhibit that compared 'healthy' and 'decadent' porcelain art.[57] Cadres wanted ceramic painters to create surface decorations based on peasant paper cuttings, calendar art, and socialist realist sculpture. In 1955, they mounted three more exhibitions to persuade artists to give up 'unhealthy' art, including public critiques of local ceramists whose wares were 'feudal', had 'unsuitable political messages' (*biaoxian bu shihe*), or 'did not reflect reality'. Clearly many local painters were reluctant to adopt the new aesthetics. Officials opened a new large collective enterprise, the Sculpture Factory, in 1956, where they directed workers to make socialist realist porcelain art.

On 18 June 1955, severe lightening cracked the skies over Jingdezhen and a downpour began. Rain continued non-stop for six days, and half of Jingdezhen flooded. Water submerged 23,693 households, 136 factories, and 230 workshops.[58] Small floods in 1951 and 1954 had led local officials to release grain stores for relief. The 1955 flood was on a massive scale – but so was the response. Officials mobilized 20,000 people to bring people, supplies, and grain to safety. All the cars, trucks, and boats in the area were used for the rescue. Eight people died in the flood, and nine were seriously injured. Rescuers salvaged a large portion of porcelain supplies. The central government sent emergency relief five days after the storm's end. Government records state that the flood kept Jingdezhen from meeting production goals for 1955, but even with this major disaster ceramists completed 77 per cent of their target.[59]

Officials were more successful at directing labour than they were at rationalizing output and coordinating distribution. Numerous government reports from the first Five Year Plan emphasized that Jingdezhen needed to produce finer wares; the new factories were not achieving the desired level of quality.[60] The city produced more than officials could distribute. Officials revised their classification system for porcelain quality and ware types multiple times. In the effort to create a smooth-functioning distribution system, the Porcelain Company repeatedly set and reset porcelain prices.

By 1957, Jingdezhen's population had grown dramatically, and the workforce was 41 per cent larger.[61] The city's standard of living was the highest in the province.[62] Private enterprise had disappeared: small private workshops had been integrated into a state-administered system, and even the street peddlers had been organized into small work teams.[63] Significant numbers of women were participating in the paid workforce. While almost everything was made with handicraft methods – only 5.4 per cent of the industry was mechanized by the end of 1957 – locals were making new tools, new types of ceramics, new pigments, decals, and gilt for surface decoration.[64] The industry earned three times as much in 1957 as it had in 1952, and brought in more foreign exchange currency than any other site in the People's Republic.[65]

## The Great Leap Forward

China's leaders wanted socialist reforms to bring about economic growth, industrialization, and modernization. They wanted workers and farmers to participate in politics, and the public to recognize the labouring masses' economic contributions. All Chinese citizens should work, and all workers should be fairly compensated. Yet central government officials disagreed about the priorities, pace, and methods of change. Their internal conflicts became starkly apparent during China's second Five Year Plan, which began with a massive push for centralization and industrialization, and ended with retrenchment, reversal, and a famine that caused the deaths of 45 million people.[66]

The idea for China to make a 'great leap forward' came from a gathering of international Communist Party leaders in Moscow to commemorate the 40th anniversary of the USSR's October Revolution. At that meeting in November 1957, the USSR's Central Committee Secretary Nikita Khrushchev announced that the Soviet Union would exceed the United States' industrial output in 15 years. Chinese leader Mao Zedong, jockeying to be recognized as the most important Communist leader in the wake of Stalin's death, responded that China would catch up with or exceed the United Kingdom in the same period.[67] In January 1958, the Chinese Communist Party announced the Great Leap Forward.

Local officials proclaimed Jingdezhen's Great Leap Forward on 2 March.[68] In May, the city government changed the traditional names of all porcelain ware types to a numerical system. In June, they founded the higher education institution that became known as the Jingdezhen Ceramics Institute, with an engineering programme and art department.[69] In July, cadres switched workers at the Number Three Porcelain Factory from piece rate to hourly wages. Soon afterwards, the workers published an article in the city's weekly newspaper (which the government had founded in 1955) stating that the entire city should change to an hourly wage.[70] Ceramists responded enthusiastically, and officials began meeting with workers to discuss implementing hourly pay.

In August, city officials received a central government directive to speed up the Great Leap Forward.[71] They immediately moved the entire workforce to hourly wages, using the average salary of 1957 as a baseline, and eliminated bonuses. Over the rest of the year, officials pushed mechanization and expansion.[72] They mechanized the china stone mines and built five new factories with technology copied from foreign countries. The new Jingdezhen Porcelain Factory was said to be the most advanced ceramics enterprise in China, capable of producing 270 million pieces of porcelain per year. The Jingdezhen Ceramic Machine Factory successfully produced a pug mill, allowing ceramists to make clay mechanically for the first time in the town's 950-year history of porcelain manufacturing.[73] Across the city, cadres replaced potters' wheels with jigger-jolly machines, and increased the numbers of coal-firing kilns. As deputy director for the Labour Bureau, Feng Shangsong helped reallocate jobs to those ceramists whose specialties became unwanted. Throwers became operators of jigger-jolly machines, and trimmers became polishers of mechanically produced bowls, plates, saucers, and cups.

Cadres urged workers to devote themselves to production and work longer hours to raise output. Tired employees hurried to do more quickly, neglecting safety, which some officials criticized as a 'rightist concern'.[74] In the Build the Nation Porcelain Factory canteen, fire broke out and spread quickly. The Labour Bureau, located adjacent to the kitchen, burned to the ground, taking with it the

Fengs' home on the second floor. Feng Shangsong was overseeing changes to production at a distant enterprise when the fire took place, leaving Mrs Feng to find accommodations for herself and their two children and salvage what she could of the family's possessions. Fires occurred frequently during the Great Leap Forward.

Across China, officials turned rural collectives into communes, and built public canteens to replace private kitchens. Demand for porcelain bowls, spoons, and plates shot up as local leaders provisioned the canteens.[75] The reorganization of farming near Jingdezhen created shortages in raw materials for porcelain production, as cadres directed labour toward food crops. The central government's Department of Energy intervened, reallocating 10,000 workers to produce china stone and pine logs for Jingdezhen.[76] Jingdezhen's porcelain factories still could not produce enough porcelain to keep up with demand.

On 20 November 1958, officials made the entire city of Jingdezhen into a single commune, establishing 84 public canteens, 106 day-cares, and 328 new factories. Cadres walked through neighbourhoods, knocking on doors to tell residents to get rid of their private kitchens and eat collectively in the canteens.[77] They called on women to leave their homes and take jobs in the factories. Thousands of women became employed as surface decorators and quality control inspectors.

The Fouliang County Communist Party formed the area's first iron-producing team on 1 December 1958. Around 13,400 workers joined production of iron and steel, some using 215 newly created backyard furnaces (see Plate 11).[78] On 1 January 1959, Jingdezhen's government ordered a number of porcelain workers and teams to quit ceramic production and make machines, clothing, watches, and other items. Officials sent other ceramists to build roads, a railroad line, and water reservoirs. In one season, 62 per cent of porcelain workers were sent to work on projects outside the ceramic factories.[79] Kiln fuel was diverted to backyard smelting, and porcelain shortages grew more acute as several factories half shut down. The Porcelain Industry Company tried to manage deficits by sending the finest wares to Beijing, Shanghai, and Tianjin, and coarse wares to the provinces.[80]

As the Great Leap Forward continued, the central government told local cadres to bring workers into management and send managers to work on the assembly line.[81] In the countryside, brigade heads were directed to do manual labour jobs, and work teams told to make production decisions. In the ceramics factories, workers-turned-managers made poor decisions about pricing, causing Jingdezhen wares to be sold at a loss in 1959 despite high demand.[82] Shortages worsened across the porcelain industry, and inflation skyrocketed.

In April 1959, some central government leaders advocated slower industrial expansion, less resources spent on infrastructure, and fewer projects. In Jingdezhen officials tried to decelerate the city's Great Leap.[83] Feng Shangsong and his colleagues at the Labour Bureau investigated the impact of the switch to hourly wages, and found that the hourly salary had decreased productivity. Workers put in their time but did not work hard. Many were producing only 70 per cent of what they had made previously. Because the factories did not count a half-day of work as a sick day, many ceramists left after half a day. Others found reasons to take breaks or run errands. Workers 'ate by the clock', as they put it.

Labour Bureau cadres switched the Red Star Porcelain Factory back to piece-rate wages with bonuses. They returned managers to management and workers to the shop floor. After a month's experiment, the city government instructed factories to correct 'excessive leftism' with regard to workers in management. Officials approved returning the entire workforce to piece-rate wages, with bonuses for workers who produced over quota. Those who exceeded targets for quantity received additional money up to 60 per cent of the value of the excess, and those who surpassed targets for quantity and quality received additional money up to 80 per cent of the value of their over-quota production.[84] Workers who under-produced had their wages cut. In September, city officials held a public meeting where they honoured 193 individuals for the quality of their work in porcelain. Thirty-two men and one woman were awarded the title of 'ceramic artist' (*taoyi meishu jia*). Others were recognized for technical expertise.[85]

Within four months, central government orders caused city officials to reverse direction. Ceramists returned to hourly wages, and cadres published propaganda declaring that piece rates and bonuses were inappropriate in a planned economy.[86] Around 10,000 porcelain workers were sent to produce steel.[87] Officials pushed the factories to adopt assembly lines, and gave each factory instructions on production specializations: the East Wind Porcelain Factory would make coffee and tea wares, the Build the Nation Porcelain Factory would make bowls and tea wares, the Red Star Porcelain Factory would make bowls and industrial porcelain, and so on. These specializations remained consistent through the mid-1990s. The government expanded the size of Jingdezhen's porcelain factories, creating nine with workforces between 2,000 and 3,000, and three with more than 5,000 workers in 1960. They increased the total state enterprise workforce to 78,225 people.[88]

Yet Jingdezhen continued to meet only 70 per cent of the nation's demand for porcelain. There were drastic shortages of bowls, spoons, and water jars.[89] The city of Shanghai wrote a letter to Jingdezhen officials, complaining about the poor-quality wares they received and the lack of spoons. In the cities of Wuhan and Nanchang, people queued to purchase porcelain, and street fights erupted. Because farmers were not producing enough straw for packing, workers used less straw to wrap porcelain for shipping. Breakages increased, further exacerbating the shortages.

In rural areas, Great Leap planning was causing food scarcity, as cadres took land out of cereal production and diverted labour to non-agricultural tasks. By the end of 1960, 10 million citizens across China had starved to death, and thousands of others died from the beatings and violence which became commonplace as cadres faced increasing pressure to achieve unrealistic goals.[90] The price of grain rose rapidly, and Jingdezhen officials lacked the funds to purchase adequate rations. Ceramic workers were shorted, and many families went hungry. Over the course of the year, a 'wandering population' of starving rural families relocated to Jingdezhen, living in makeshift tents along the banks of the Chang River.[91] A kerosene lamp started a serious fire in the camp on 1 October 1960, burning to death 107 people, injuring 61 others, and damaging surrounding property.

The central government began an 'adjustment period' in 1961. Cities were told to reduce the size of urban population and cut the workforce, and the central government decreased investment in industry, shutting down tens of thousands of state enterprises.[92] Jingdezhen officials reduced the number of urban residents by 8,688 people, recording that 3,000 'volunteered' to leave and return to farming. They instructed factories to lay off workers: 10,987 in 1961, 10,225 in 1962, and 8,938 in 1963, for a total of 30,150 people.[93] Officials closed day-cares, schools, and barbershops inside the state enterprises to reduce the workforce. Women and rural residents were sent home. While Mrs Jiang kept her job, many of her female neighbours left the factories, and some never returned to work. Jingdezhen's Central Party Committee split several factories into two: the Red Star Porcelain Factory, the Red Flag Porcelain Factory, and the China Electric Porcelain Factory. Several state factories were turned into collectives, and many small collective factories were closed. A number of factories returned to handicraft production.[94] Officials adjusted the porcelain production quota down to 500,000 pieces, Jingdezhen's lowest level of output since 1955.[95]

In 1962, central government leaders told city officials to reform the state enterprises. Cadres began experimenting at the Universe Porcelain Factory.[96] They allocated the factory a fixed number of employees, of whom 89 per cent had to be directly involved in production. Officials determined how much clay, glaze stone, overglaze enamels, and transfers should be used to produce a fixed volume of finished porcelain. They returned workers to piece-rate wages, allocated according to the quantity and difficulty of the wares produced. They introduced mechanical glazing machines, increased the numbers of jigger-jolly machines, and added more mechanical assembly lines. Outside the Universe Porcelain Factory, officials established a work team to build coal kilns.[97] By the end of 1962, the team had constructed 166 new tunnel kilns, and officials declared that the porcelain industry had switched to coal. Ten wood-burning kilns remained intact, at Build the Nation Porcelain Factory, the People's Porcelain Factory, and the Art Porcelain Factory, to be used for special orders and educational purposes.

Jingdezhen's second Five Year Plan saw significant inflation, inadequate supplies, poor decisions about output, and misuse of labour on futile projects, such as steel and iron production. Women were mobilized to join the workforce and then laid off. Ceramists underwent changes in compensation multiple times, and residents experienced their first rationing shortages. Yet Jingdezhen's Great Leap was mild compared to the deaths, violence, and damage seen in other parts of China. While older ceramists remembered the gap between what the government planned and what the factories actually did, they characterized the Great Leap Forward as the foundation for all future development in Jingdezhen's porcelain industry. The technologies adopted, including jigger-jolly machines, assembly lines, transfers and decals, and coal kilns, were used into the early 1990s, and in some cases, continued in use through 2010. Productive capacity increased. During the push to modernize, city officials paved roads, added factories, and built a major water reservoir. The Jingdezhen Ceramics Institute, founded in 1958, is still China's premier site for higher education in ceramics.

## Economics or politics?

From 1961 until 1966, Premier Zhou Enlai and his Finance and Economics Small Group shrunk investment in industry and construction, reduced the size of the state sector, dismantled communes, and promoted more local decision-making.[98] These policies helped the economy to recover and reduced the enormous budget deficits that the nation had accrued during the Great Leap Forward, but caused social problems. In urban areas, including Jingdezhen, a significant number of people were laid off. Young people who were old enough to work waited at home for the Labour Bureau to find them a job. Some began to feel disillusioned with the New China.[99]

In an effort to mitigate unemployment, from 1961 to 1966 the government allowed limited private commerce. In the villages around Jingdezhen, farmers got small private plots, and some rural markets reopened. Urban peddlers resumed private sales and city officials opened small markets for 'off-plan' commerce.[100] The government set up six small collective factories for surface decoration and the production of electrical ceramics, and rehired some of the residents who had been sent home. Of the 30,150 workers who were laid off between 1961 and 1963, cadres found work for 13,162 between 1963 and 1965.[101] Cadres also encouraged youth who had completed school to 'go down to the countryside' and build the 'new socialist village' and 'new socialist farmers'. The first 100 youth left Jingdezhen on 30 June 1964, after a large official send-off. Over the next year, 2,308 more followed.[102]

At the local, regional, and national levels, government leaders agreed that Jingdezhen should increase production of porcelain. Output for 1961 and 1962 was far lower than the industry was capable of. Cadres procured funds from the regional and national governments to build factories that produced technology for the porcelain industry: electric meters for kilns, ceramics production machines, printers for transfers, even a two-headed jigger-jolly machine that officials decided would become the industry standard.[103]

Production increased sharply between 1963 and 1966 – too sharply. In 1964, only 61 per cent of the total output passed quality inspection. In September, the Porcelain Industry Company wrote a letter of complaint to the Red Flag Porcelain Factory, detailing its problems: workers were not finishing wares properly, skimping on glaze, allowing flawed pots to move down the production line, and producing pots with glaze shrinkage.[104] At the same time, the city was manufacturing too many mid- and high-end wares when domestic buyers wanted cheaper porcelain. Ceramic industries in Hebei, Hunan, and elsewhere, which officials had developed during the Great Leap Forward when Jingdezhen could not keep up with demand, sold inexpensive ceramics. Jingdezhen's factories began stockpiling large quantities of wares, putting from 60 to 220 per cent more porcelain in storage each year.[105] The Porcelain Industry Company officials adjusted prices eight times between 1961 and 1965, hoping to recapture the domestic market. Jingdezhen did manage to hold on to its status as the number one exporter of ceramics in the country, with markets in Singapore, Malaysia, Egypt, Africa, the United Kingdom, Sweden, and Holland.[106]

Urban construction slowed after the Great Leap Forward, but on 1 October 1963, officials hailed a major achievement: the city's first permanent bridge.[107] A permanent bridge was the precondition for large-scale future construction on the western banks of the Chang River. Government records describe many city residents volunteering to help build the structure, and construction took only one year. Local ferry transportation ceased, leaving Labour Bureau officials with more residents for whom they needed to find jobs.

Late in 1962, Mao Zedong announced that China had forgotten the importance of class struggle. Mao saw the steps taken to normalize the economy as a movement toward capitalism. He launched the 'Four Cleanups' campaign, also known as the Socialist Education Movement, to purify the Party. City officials in Jingdezhen began an anti-corruption campaign among party cadres in April 1963. Seven months later, they brought the Four Cleanups to five rural communes. The response was tepid. Just over half of the work teams complied with the regulations to struggle against 'enemies of the people'.[108]

## The porcelain capital's Cultural Revolution

The fears that caused Mao to initiate the Four Cleanups did not dissipate. Mao's wife Jiang Qing and some other party members argued that artists and authors, some of whom held political office, were publishing slightly disguised criticisms of the Party. Mao and his ally Lin Biao denounced and ousted some top political leaders. The Jingdezhen newspaper began reporting the criticisms of prominent Party members on 12 May 1966.[109] On 16 May, Mao and his allies announced that capitalists had infiltrated the Party, including at high levels. Shortly thereafter, a philosophy lecturer at Beijing University posted a 'big character poster' (a large banner written in big calligraphy), denouncing the university leadership. When Mao endorsed the poster, students across the countryside began revolting against school principals, party secretaries, and local officials.[110] Students at Jingdezhen's Number Two Middle School posted the city's first big character poster on 15 June 1966.[111] They took to the streets and denounced a barbershop, restaurant, and pharmacy as capitalist, forcing proprietors to take down their shop signs.

Municipal officials responded by trying to manage the movement, just as the central party leaders who promoted the post-Great Leap economic adjustment tried to direct the new 'Cultural Revolution' at the centre. On 25 June, local officials set up work teams at the Jingdezhen Ceramics Institute, the Light Industry Bureau Ceramics Research Center, four middle schools, the newspaper, and the city dance troupe to organize Cultural Revolution activities. Upon receipt of a central government dictate to 'clean out' all capitalists and capitalist policies in the porcelain industry on 2 August, officials switched ceramics workers back to hourly wages and cancelled bonuses.[112] One month later, the Porcelain Industry Company 'revolutionized' all 'feudal, capitalist, and revisionist' ware types, surface decorations, brand names, factory names, and commercial marks. One hundred porcelain patterns and 150 kinds of sculpture were taken out of production. Officials closed the Ceramics Museum, turning it into a porcelain factory. In the state and collective enterprises, ceramists held criticism sessions and wrote big character posters, producing more than 50,000 big character posters in one month.[113]

Jingdezhen's young people refused to let the government lead the attack on hidden capitalists. On 23 August, students from the Jingdezhen Ceramics Institute, middle schools, high schools, and vocational schools went on the hunt, attacking families with bad class labels and ransacking their homes for any signs of capitalism, revisionism, or counter-revolutionary materials.[114] They smashed antique porcelain and traditional sculpture, and ripped up old books and foreign materials. Ten days later, 1,000 students and teachers founded the city's first 'Red Guard' organization of loyal 'soldiers' following the 'Red Commander' Mao Zedong. They put up big character posters at the city government, calling for the removal of top leaders. A few days later, Red Guards began a public criticism session at the city government offices that lasted for two days. They denounced the city officials as capitalists and revisionists, and began bringing leaders out to be struggled against. Some cadres and residents tried to stop them, and the protest deteriorated into a brawl. Ten days later, several hundred students and teachers formed an alternative Red Guard organization. Miners in the china stone quarries and local farmers followed with their own Red Guard groups.

Twenty-one-year-old Sun Fuji was a glazer at the state-owned Radiant Porcelain Factory (*Guangming Cichang*) when the students began the Cultural Revolution. Curiosity carried him out of the factory to see the protests. He thought 'rebelling' (*zaofan*) looked fun and quickly joined a Red Guard group, hanging big character posters and breaking porcelain decorated with dragons and phoenixes. He and his fellow Red Guards went around to the neighbours' houses looking for the gold jewellery that women traditionally brought for their dowries; some of the families threw their gold into the latrines so that the Red Guards would leave them alone. Sun and his companions organized struggle sessions against the cadres in his factory, and then turned against officials in the city government. Sun was delighted that he still got paid even though he wasn't

doing any glazing. When some university students left for Beijing, hoping to see Chairman Mao, Sun accompanied them to the provincial capital. It was his first trip out of Jingdezhen. Since he hadn't told his parents he was leaving, and Nanchang was cold, he decided to return to Jingdezhen. Somehow on the way he managed to get a gun – 'from the army', was all he would tell me.

On 14 October 1966, city officials publicized five suggestions.[115] 'The people of Jingdezhen have many different ideas about the Cultural Revolution,' they wrote. Cadres urged everyone to stop making big character posters, since shreds of posters were littering the streets because competing Red Guard factions ripped down one another's signs. They said street protests were blocking traffic and impeding production and should cease. Red Guards seized on the suggestions as evidence that the city officials wanted to halt the Cultural Revolution. A group set out for the provincial capital to demand the removal of Jingdezhen's leaders. En route they clashed with farmers, and many were injured. Another group held a massive street protest, which also turned into a brawl. One Red Guard faction broke into the city's central rationing office for grain. Other Red Guards, some armed with guns, attacked them. Several people were injured and property destroyed. On 20 January 1967, a Red Guard faction convened a crowd of 20,000 on the People's Square and declared that they would smash capitalism.[116] Four days later, Red Guards forced the city government and Central Party Committee to step down.

Feng Shangsong was among those removed from their posts. Red Guards pulled Feng from his office and insisted that he admit his 'capitalist' ways. He and other former city leaders were made to wear tall white hats and carry broomsticks and were paraded through the streets. Students, workers, and other residents hit and spat on them. Red Guards then gave Feng work as a janitor, ordering him to write daily self-criticisms and read them aloud in front of Chairman Mao's portrait. At any time, groups of youths would stop him, gather a crowd, and struggle against him.

A Red Guard faction formed a Revolutionary Management Committee and took over the Porcelain Industry Company and other aspects of city government.[117] They ordered the immediate and complete cessation of all private commerce. Small collective factories were joined into large collectives. Large collectives were treated as state enterprises. Since the new Revolutionary Management Committee had little experience with governance, there were shortages of china stone and coal. Work stopped frequently for political sessions, group recitation from Mao's Little Red Book, and criticism of former managers and cadres. Ceramists smashed porcelain in storage, declaring it capitalist or feudal.

Over the course of 1967, Red Guard factions held numerous protests, many of which ended in violence.[118] Schools were effectively closed. In April, Red Guards from the Number One Middle School staged a week-long sit-in at the Post and Telecommunications office, disrupting mail, traffic, and news reports. In September, another Red Guard group protested for three days at the People's Bank, demanding that bonus pay be reinstated for porcelain workers. They blocked traffic and damaged power lines. Days later, two Red Guard factions fought one another with guns and other weapons at the electrical plant. Sun and his Red Guard faction built a fortress on Dai Family Alley, a large round structure with holes for guns, where they could fire at 'the enemy' while remaining protected inside. On 22 November, two Red Guard factions had a bloody three-hour shootout that only ended with the intervention of Army Unit 6300. Sporadic conflicts continued between Red Guard groups armed with steel poles, knives, guns, and spears until the middle of 1968.[119]

As Lunar New Year approached in January 1968, students rounded up Feng Shangsong and several other former officials, including Jingdezhen's former mayor, and imprisoned them in an elementary school. The prisoners were denied food or water, but the mayor's wife managed to smuggle in some steamed buns on Lunar New Year's Eve. On Lunar New Year's Day, Red Guards marched the men outside for a public criticism session, but Feng and his colleagues refused to admit any wrongdoing. Red Guards cursed them, beat them, and hung signs around their necks proclaiming them traitors. From then on, these former leaders were isolated except for re-education sessions. In May, Red Guards sent them to the countryside for rehabilitation through manual labour.

On 23 April 1968, Red Guards gathered 80,000 residents and announced a third campaign against capitalism.[120] In the following weeks, Red Guards sent 741 factory managers and technicians to manual labour jobs in the countryside and 2,100 artisans to join the army. At the Ceramics Research Center, Red Guards declared 157 scientists and technicians counter-revolutionaries. They convened a struggle session, during which the male head of the centre and a female translator were killed. Cadres, scientists, and skilled workers were removed from their posts. At the Art Porcelain Factory, 75 employees and their families were made to go to the countryside to perform manual labour (Table 3). Seeing people with good class labels become political targets, many ceramists began to feel unsafe. Workers accused their colleagues of being capitalists or counter-revolutionaries in the hopes of deflecting violence away from themselves and their families.

**Table 3** The Cultural Revolution's impact on labour and output in the porcelain industry, 1960–1972[121]

| | 1960 | 1965 | 1968 | 1971 |
|---|---|---|---|---|
| Number of technical workers | 34,000+ | | | 24,297 |
| Skilled colour glaze decorators at Build the Nation Porcelain Factory | | 180 | | 29 |
| Skilled famille rose decorators at Art Porcelain Factory | | 986 | | 140 |
| Ware types | | 1,700+ | | 300+ |
| Rate of exports | | | | 21% less than 1965 |

In July 1968, the Red Guard Revolutionary Management Committee took steps to de-escalate the campaign.[122] On 23 July, they organized a graduation for 4,618 middle school, high school, and university students, and two days later notified the new graduates that they should become workers or farmers. With the assistance of the army, the former students were sent to china stone mines, farms under army control, and the far northwest province of Xinjiang. On 5 September, the Revolutionary Management Committee sent another 2,400 students and 120 cadres to the countryside, again using army troops to enforce the removal. Over the next several years, thousands more youth were sent to the countryside – as many as 17,837, according to one source.[123]

Violence diminished significantly, but as Lin Biao initiated a new campaign to promote militarization across Chinese industries late in 1968, other problems emerged. On 28 December 1968, the Jiangxi Provincial Revolutionary Committee ordered Jingdezhen to disband the Jingdezhen Ceramics Institute and the Light Industry Bureau Ceramics Research Center.[124] The Revolutionary Management Committee sent activists to tear apart classrooms, libraries, and other facilities, proclaiming that 'porcelain cannot be used to fight a war', and 'porcelain cannot be used as a bomb'.[125] Ceramics factories were prohibited from repairing tools or machines, adding jobs, or adjusting supplies. Over the next 18 months, the Revolutionary Management Committee turned seven state-owned porcelain factories to other forms of production, including munitions, cars, broadcast equipment, electrical supplies, screws, and textiles. Around 1,600 miners were taken out of the china stone mines and made into a propaganda team. The Jingdezhen Porcelain Factory, heralded in 1957 as China's most advanced ceramics enterprise, was disbanded and given to the Defence Department, which turned it into the Chang He Airplane Factory. In the remaining porcelain factories, workers experimented with making ceramics more like iron. By the end of 1970, Jingdezhen had slipped from China's 1st to 4th largest export site for ceramics.[126]

In February 1970, Jingdezhen's Revolutionary Committee received a central government directive to attack counter-revolutionaries, capitalists, and thieves.[127] At the Chemical Ceramics Plant, 105 technical workers were labelled 'ox ghosts and snake demons', a term from ancient poetry used at this time to refer to nebulous political crimes. Red Guards forced the workers to wear tall white hats and signs and struggled against them in public. At the Red Flag Porcelain Factory, two-thirds of the workforce was criticized. Twenty per cent of party members at the East Wind Porcelain Factory were forced to leave Jingdezhen. In September, the Revolutionary Committee sent 12,500 'capitalists, class enemies, and rightists' to the country for reform through manual labour.

Feng Shangsong was allowed to return to Jingdezhen in May 1970 and resume work at the Labour Bureau. He was philosophical about his experiences. 'When I lived in the countryside, I tried to do good works,' he told me. 'Anyway, from this time forward the Cultural Revolution became more orderly.' Jingdezhen's Revolutionary Management Committee had begun to review the cases of former officials, technical workers, and scientists who had been accused of political crimes. By December, 1,500 cadres had been rehabilitated and returned to their positions.[128] In February of 1971, the central government declared that all institutions of higher education and universities should reopen. Jingdezhen leaders selected 34 farmers, soldiers, and porcelain workers to attend universities in Nanchang and Beijing.[129] A year later, in February 1972, China had recovered its United Nations seat and initiated diplomatic relations with the United States. In Jingdezhen, the Porcelain Industry Company resumed management of clay and fuel, porcelain transportation, and the factories.[130] Jingdezhen's primary and middle schools began teaching again. In July, the Revolutionary Committee reopened the Ceramics Research Center, and returned the Red Flag, Build the Nation, and Art Porcelain factories to ceramics production. In August, government leaders designated several areas in Jingdezhen for private commerce, including Dai Family Alley, where Sun and his compatriots had built their fortress four years before. Rehabilitation of cadres continued, with 1,145 cleared of crimes by September of 1972. The ceramics industry was functioning normally, although some of the factories that had been taken out of ceramics production were never returned to porcelain manufacture. On 4 October, the central government declared Jingdezhen an 'open city' (*kaifang chengshi*), meaning that the city could host foreign visitors.

Jingdezhen's Cultural Revolution set the porcelain industry back, wreaked havoc in the lives of officials and experts, and caused the destruction of many culturally valuable objects. Yet relative to other places, the town's Cultural Revolution was shorter, less damaging, and less deadly. The majority of Jingdezhen's ceramists had good class labels, little education, and never served in management or government, making them unlikely political targets. Many adults who lived through the Cultural Revolution described it as a time when they, their neighbours, and local officials lived morally exemplary lives.

Jingdezhen's ceramic industry was bigger, more stable, and better managed in 1972 than it had been in 1949. The scope of wares Jingdezhen produced had diversified, the number of people employed in porcelain manufacturing had increased dramatically, and more women were in the workforce. The city's infrastructure had been strengthened and expanded, and residents' standards of living were higher. On the balance, big government's return to Jingdezhen had improved ceramists' lives.

# 5

# Dual-Track Porcelain, 1973–1993

On 7 December 1974, three Americans who worked for Boehm Porcelain of Trenton, NJ, arrived in Jingdezhen. Leading the group was Helen Boehm, the American 'Princess of Porcelain' and company head. Maurice Eyeington, Boehm's head artist, and Frank Cosentino, her executive assistant, accompanied her. Boehm Porcelain had sculpted the pair of life-sized porcelain swans that President Richard Nixon gave Chairman Mao Zedong when the United States and the People's Republic of China re-established political relations in 1972. In 1973, Helen Boehm helped host an official Chinese delegation in Washington, DC. Shortly thereafter, the PRC department charged with managing US–China relations issued her an invitation to Jingdezhen. She and her staff were the first Americans to visit the porcelain capital since the 1920s, and only the second group of westerners since the 1952 expulsion of foreign missionaries (a group of ceramists from New Zealand had just visited in November). Aware that Americans were eager to catch a glimpse of the People's Republic, Cosentino wrote an account of their trip that Boehm Porcelain published in 1976.[1]

Jingdezhen seemed isolated, provincial, and thoroughly polluted with coal dust to the Americans. Residents dressed in Mao suits and lived in 'old and very modest' houses. Many roads were unpaved and the trio saw few cars. Transport was manual. 'Carts carry everything, including clays and porcelains,' wrote Cosentino. 'Men move from building to building at double-time speed while balancing a rack or board of delicate porcelains on one hand held above the shoulder.'[2] Most porcelain was shipped out on the river, although officials were building a railroad to Nanchang.

Yet while the town was a backwater, its ceramics industry was impressive.[3] The delegation's first stop was the For the People Porcelain Factory (*Weimin Cichang*), which mass-produced bowls, plates, tea and coffee sets, and vases. The wares were high quality, with 70 per cent destined for export. Manufacture was mechanized, and included 'innovative' technologies that the three had never seen before: double-headed jigger-jolly machines for throwing, foot-pedalled mechanized glazers, and a 285-foot-long coal tunnel kiln. The potter operating the double-headed jigger-jolly used both hands to shape forms with speed and coordination. When the glazer pushed her foot pedal, liquid glaze shot up from a large tub for a split second, glazing wares more efficiently than the Boehm's sprayers (see Plate 12). The kiln fired greenware stacked on carts that rotated continuously through the tunnel.

The factory's deputy director and party secretary told the Americans that For the People's workers had developed 147 technical innovations that revolutionized the factory's production system. In turn, For the People took good care of its 2,000 employees. When a worker had a baby, she received a special allowance for day care. Her child was entitled to free schooling. She could retire at age 50, and her husband could retire at 55. Her pension was as high as 90 per cent of her wages, if she had worked at the factory long enough. Upon her death, the factory paid for funeral expenses.

Next was a visit to the Art Porcelain Factory (*Yishu Cichang*). Most of the factory had been diverted to manufacturing electronics and broadcast equipment in 1970, but officials had returned it to porcelain production in July 1972.[4] The American delegation raved about the factory's 900 ceramic painters. 'These painters are the equal of the best in the world, and there are hundreds of them,' exclaimed Cosentino. The trio marvelled at the extraordinary craftsmanship of hand-glazed black vases decorated with polychrome overglaze enamels, an enormous porcelain tile painting of Mao as a young man, and two eggshell-thin lamps adorned with butterflies and fish. 'It is apparent that this quality of work can only be done in a society where pay and costs have little relationship to the finished product,' Cosentino reflected. Given what he saw at the Art Porcelain Factory, Cosentino continued, '[T]here is no question that with their excellent artistry and quality and low costs, the Chinese eventually could dominate the world markets in all of their handicrafts and arts, especially in fine porcelain.' The only deficiency was too little gold in the gilt. The Boehm delegation felt the producers over-economized, yielding a 'thin,' 'brassy' colour.

At the Sculpture Factory, the trio again found 'outstanding' talent, fine workmanship, and high-quality wares, although most of the sculptures were too political for their taste. Still, sculptors made models 'that compare with the finest work': a pair of peacocks with flowers, a maiden training ten deer leaping and running about her, delicate spider chrysanthemums with foliage mounted on a plate. Almost nothing was made by slip-casting, the dominant method at Boehm Porcelain. Instead, sculptors used press moulds to form figures, followed by extensive tooling and finishing.

Other sites that the Boehm group visited included the Ceramics Museum and a ceramics research institute.[5] The Ceramics Museum was disappointing, with too few examples of Jingdezhen's finest imperial porcelain and an overabundance of 'politically oriented' pieces. The institute was 'impressive and modern', with spectrographic, chemical, and thermal technology imported from the United States, Germany, Japan, and Switzerland, and a 'professional' staff creating new designs, analysing materials, and studying production. Part of the institute was a manufacturing facility where the Americans met skilful hand-throwers and sculptors who persuaded Eyeington to demonstrate his modelling techniques. Later, at a meeting attended by designers and technicians from 'all the important factories', Boehm, Eyeington, and Cosentino described in detail the production methods at Boehm Porcelain, to reciprocate their hosts' openness.

The trio left believing that Jingdezhen was positioned to dominate global ceramics. 'Their artists and craftsmen are the equal of the world's best,' wrote Cosentino, '[T]heir thin, white porcelain is the world's finest.'[6] Jingdezhen's international market was booming. Levels of employment were high; to Cosentino, the industry had 'an exaggerated division of labour which puts everyone to work who is able'.[7]

One of the painters whom the Boehm delegation might have seen at the Sculpture Factory was Liu Changshan. Liu started work at the Sculpture Factory in 1954 at the age of 20. Prior to that he had painted with his father in the family's tiny workshop. In the early 1950s his father had joined a painting cooperative, which became part of the Sculpture Factory, a large collective run by the city government. Liu and his father had good class labels because the family had been poor. Liu described himself as an orphan, because when Liu was a baby, his biological father had staked him in a game of mahjong and lost. Liu grew up with his adoptive father. He had a deep hatred of gambling.

The government determined how much the Sculpture Factory should produce based on the enterprise's total number of employees producing at a constant rate over eight hours for 25.5 days each month. Liu worked an eight-hour day, six days per week. He and his co-workers had seven paid holidays per year: two days at National Day, three days at Lunar New Year, one day for Worker's Day, and one day for New Year's Day. Liu remembered that food was cheap, prices were stable, and cadres were honest. 'Political movements kept the officials and managers in line,' he said. Liu found it satisfying to participate in the factory's criticism sessions.

Sculpture Factory ceramists could not always meet the production levels that the government planned in the time allotted. Beginning in the late 1970s, the collective offered Liu opportunities for overtime. He could stay late after his shift (*jiadian*), or take an extra shift (*jiaban*). The factory was supposed to request permission to offer overtime, but managers often ignored this rule. As a painter, Liu was especially in demand for extra shifts and hours. City officials tried to prevent the ceramists from working late by locking the factory doors after the official end of the workday, but Liu and his co-workers climbed in through the windows. The government cut electricity to the factory after work hours, so the painters brought their own kerosene lamps and candles.

Some of Liu's overtime compensation came in the form of extra grain-ration coupons. He earned so many that he, his wife, and his four children could not eat all the food. They exchanged some of the coupons for eggs, and sold others on the black market. One day when the Lius were reminiscing about the 1970s and 1980s, Mrs Liu brought out a large tin container to show me. Inside were dozens of ration coupons that the family had never redeemed.

City officials and ceramists working in state and collective enterprises called the 1970s through the early 1990s a 'golden age'. They believed that Jingdezhen's porcelain industry reached a higher level than that attained in the Ming or Qing dynasties during these two decades. The government invested substantially in porcelain manufacturing, building new factories and importing new technologies. Employment expanded, and city leaders reinstated bonuses and honorifics. Wages and compensation were secure, and there were no setbacks from political campaigns. After the City Revolutionary Committee announced the central government's decision to permit private commerce on 31 December 1980, new opportunities emerged in the private sector. Jingdezhen's first private porcelain enterprises opened in 1983, and more followed after China's 13th Party Congress recognized private enterprise as a necessary part of the economy in 1987.[8] Yet most workers in Jingdezhen's state and collective enterprises were not willing to quit their secure jobs to head private firms or work for private entrepreneurs. Instead, many potted and decorated in private workshops during their off hours, bringing with them pilfered supplies of clay, pigments, and transfers from their work units.

## Reviving the porcelain industry

Jingdezhen's Revolutionary Management Committee accelerated efforts to normalize porcelain production and other aspects of daily life after the central government declared Jingdezhen an open city. Between 1973 and 1976, officials allowed porcelain experts, skilled technicians, factory managers, ceramists accused of being capitalists, and sent-down youth to come back to Jingdezhen. In 1973, more than 500 experienced managers, 4,000 collective workers, 333 petty entrepreneurs, and 3,914 young people returned.[9] Many went back to their original jobs in porcelain enterprises, but since some of the factories had been diverted to other industries, some returnees were given positions at new work units. Youth who had not been employed before going to the countryside were given jobs in porcelain factories or county institutions. Thousands more sent-down youth came home subsequently: 2,564 in 1975; 7,898 in 1976. In January 1979, the Revolutionary Management Committee began rectifying citizens' records to remove 'erroneous' classifications such as enemy of the people, counter-revolutionary, and capitalist.[10]

Between 1973 and 1978, officials restored the city's ceramics vocational schools, the Jingdezhen Ceramics Institute, and porcelain research centres.[11] Two ceramics training schools and the design facilities of the Jingdezhen Ceramics Institute reopened in April 1973. The Light Industry Bureau Ceramics Research Center began functioning again the following year. In December 1975, the central government ordered that the entire Jingdezhen Ceramics Institute should reopen; the Revolutionary Management Committee achieved this on 2 August 1976. In 1978, six state porcelain factories recovered their middle and high schools.

Jingdezhen's open status meant that foreign delegates could visit, factories could participate in international ceramics competitions, and enterprises could produce for global markets.[12] The first visitors came from Japan for technical and artistic exchange and the purchase of porcelain in April 1973. Porcelain buyers from Hong Kong were permitted to travel to Jingdezhen later that year. Next came the New Zealand ceramists in November 1974, followed by the Boehm delegation in December. Six months later, in July 1975, a Romanian group of ceramists visited, and in September, a delegation of Australian ceramists. More and more foreigners came to Jingdezhen in subsequent years, with 278 international guests visiting in 1978. In April 1979, city officials hosted a ceramics fair to give outside buyers the opportunity to meet directly with representatives from porcelain factories.

Jingdezhen's porcelain enterprises competed in international competitions in 1974, 1976, 1977, 1979, 1980, and 1984, winning three international prizes.[13] The People's Porcelain Factory (*Renmin Cichang*) was especially successful. Small numbers of factory employees were allowed to travel abroad to participate and see first-hand the kinds of porcelain that international companies were producing. In 1980, city officials permitted several factories to set up exhibitions of porcelain in Hong Kong.[14] This increased visibility resulted in new business: a large order from 27 Hong Kong merchants in 1975, an order by the Japanese porcelain company Mikasa in 1979, a contract with an American company in 1980. By 1979, Jingdezhen had returned to its status as the top earner of foreign currency in China.[15]

China's highest officials stressed the importance of Jingdezhen's porcelain industry. On 7 February 1973, the newly rehabilitated Deng Xiaoping, his wife, and five other central government leaders arrived in Jingdezhen to tour porcelain manufacturing sites. Over five days, they visited the Ceramics Museum and eight state and collective factories: the Art Porcelain Factory, the Sculpture Factory, For the People Porcelain Factory, Radiant Porcelain Factory, the People's Porcelain Factory, Build the Nation Porcelain Factory, Red Flag Porcelain Factory (*Hongqi Cichang*), and the Universe Porcelain Factory.[16] The leaders lauded the industry's size, comprehensiveness, and mechanization. Yet Deng Xiaoping also voiced concern that Jingdezhen's traditional handicraft manufacturing had almost disappeared.

On 25 April 1973, officials declared a Five Dynasties kiln site in Hutian, a village on the outskirts of the city, a public preservation area, and prohibited anyone from digging at or harming the site.[17] The Porcelain Industry Company returned traditional wares to production, including porcelain decorated with or in the shape of figures from traditional Chinese religions: the Eight Immortals, Guanyin, Luohans (Arhats), the Gods of Longevity, Prosperity, and Officialdom, dragons, phoenixes, and so on. In November 1974, cadres allowed factories to manufacture 31 types of traditional blue and white wares.[18]

When Deng emerged as China's top leader after Mao's death in 1976, Jingdezhen's officials implemented two other preservation initiatives. The first was construction of the 'Ancient Kiln and Folk Custom Museum' (*Guyaominsubolanqu*). Cadres ordered that the remnants of several old porcelain workshops be moved west of the Chang River and reassembled into a single traditional-style comprehensive workshop, with two wood-burning kilns. Well over a hundred ceramists were brought out of retirement to work at the Ancient Kiln, using stick-spun potter's wheels and other old tools to show handicraft production methods. Cadres formally opened the Ancient Kiln Folk Customs Museum on 6 August 1980 (see Plate 13).[19]

The second initiative was a training programme for the children of high-level cadres.[20] In 1977, officials assigned each government office a quota to fill of participating children. Highly positioned ceramic artists were the first to select a son or daughter to take part, followed by other officials. Two hundred families sent boys and girls between the ages of 14 and 16 to apprentice with skilled older ceramists at the Jingdezhen Ceramics Institute and the Light Industry Bureau Ceramics Research Center in handicraft methods. The masters received a special salary and other kinds of favourable treatment in recompense. Participants recalled that they were happy to be selected as designates. 'I was joining the workforce and would get a salary,' one woman told me. 'Back then, everyone thought working was more important than education. Don't forget, 1977 was before the national college entrance examinations were reinstated.'

City officials began reinvesting in Jingdezhen's porcelain industry in 1976. They authorized the construction of 16 tunnel kilns and the opening of two china stone mines. They permitted street committees to establish 15 collective porcelain factories employing a total of 8,490 ceramists. Officials supported research to design a fully automated porcelain manufacturing system (the effort ultimately failed).[21] Over the next year, the government added three more collective porcelain factories and two factories producing packaging for ceramics.[22]

## Deng Xiaoping's 'Reform and Opening'

At the time of Mao's death, China had a stagnant economy, declining agricultural outputs, and a low standard of living.[23] The People's Republic had few allies or trading partners since relations with the Soviet Union and Eastern bloc had soured in 1958. Deng Xiaoping was among those leaders who looked to private commerce, foreign trade, and foreign investment to jumpstart development. After he emerged as China's paramount leader in mid-1977, Deng adopted the policies known as 'Reform and Opening' (*gaige kaifang*). Between 1978 and 1981, Deng implemented the 'household responsibility system' in agriculture, encouraging farmers who met their quotas for agricultural products to sell their surplus at private markets and keep the profits. He opened the People's Republic to foreign direct investment, establishing special economic zones in Zhuhai, Xiamen, Shenzhen, and Shantou to attract foreign firms. He allowed private entrepreneurs, called *getihu*, to own and operate businesses that produced entirely for the market. The bulk of China's economy remained on the plan, and the government continued to fix prices for manufactured commodities, but Deng kept factory quotas flat. He allowed state and collective enterprises to sell their surplus on the market, creating a system of dual-track pricing (government-fixed and market prices).[24] Deng also saw slowing population growth as essential to China's economic wellbeing, so he implemented the One Child Policy in 1979.

Rural incomes and international trade increased quickly under the new measures. Demand for Jingdezhen porcelain surpassed what the industry could produce.[25] City officials sought central government investment to expand, and received an average of five million yuan per year between 1979 and 1984.[26] In 1978, city officials successfully applied to build a new state enterprise, the China Wind Porcelain Factory. Construction began in 1982, and the factory opened in 1987. China Wind boasted a state-of-the-art facility and introduced a revolutionary practice to Jingdezhen: bisque firing.[27] Throughout Jingdezhen's history, porcelain had been thrown, trimmed, decorated, and then fired once (sometimes called raw firing). In China Wind, porcelain would be thrown, trimmed, and fired to become a 'bisque' or 'biscuit', and then decorated and re-fired, as was typical in North American and European ceramic production. In 1979, cadres spent 8,000 yuan to build a handicraft collective and hired 500 ceramists to produce blue and white porcelain desk sets. Over the following months, officials funded a team of experts from Jingdezhen and Shaanxi Province to build a mechanized press for architectural tiles. The team succeeded in making China's first clay dust press on 7 August 1980. Next the city government began building a tile factory: the Jingdezhen City Architectural Porcelain Factory, equipped with the new dust press, opened in 1982.[28] Officials decided to build a small coke factory (coke is a fuel made from coal used in blast furnaces), hoping that they could pipe the coal gas (a by-product of coke manufacture) to the porcelain industry's kilns for fuel. When the experimental factory succeeded, city leaders appealed to the central government to fund a larger plant. They received 53 million yuan for the project, which was completed in December 1986.[29]

In 1982, the State Council of China announced its designation of historical and cultural landmarks across the country. Jingdezhen was awarded national status as a historic place for its thirty-plus historic kiln sites dating from the Five Dynasties period through the Qing dynasty.[30] In March 1982, the central government awarded Hutian national recognition for its ancient kiln site.[31] Archaeologists from Beijing joined faculty from the Jingdezhen Ceramics Institute to excavate and do research. During a bad storm on 20 June, 34 were killed and three severely injured in a bus accident by the Chang River.[32]

In the porcelain factories, cadres experimented with incentives to increase productivity. Officials switched a few collective factories to piece-rate wages in 1980. Any worker who produced over quota received a bonus based on the piece-rate pay scale. The new wage structure must have worked, because in 1981, the government changed 9,870 more ceramists to piece-rate wages.[33] City officials also increased the size of bonuses for state and large collective workers (90 per cent of whom remained on hourly wages), allowing bonuses to reach the equivalent of a month's salary.[34] They told the porcelain factories to sell their surplus on the private market if they produced over quota.[35] Between 1979 and 1984, porcelain output increased by an average of 6.5 per cent per year.[36]

In 1978, according to official records, not one person worked in private enterprise.[37] Jingdezhen had a total of 180,400 urban workers, of whom 141,100 worked at state enterprises and 39,300 worked at collectives. The city's ten large state-owned porcelain factories each employed between 2,000 and 5,000 workers, and the four city-run large collectives each employed between 900 and 2,000 workers. Ceramists at state and large collective porcelain enterprises had permanent employment with stable wages, comprehensive benefits, and retirement pensions: 'iron rice bowls', as workers put it. Porcelain workers were proud, even smug, about their job status, and local families told their daughters to marry a state enterprise porcelain factory worker. 'Getting a job at a [state enterprise or large collective] porcelain factory was harder than getting into heaven (*jin cichang bi jin tiantang yao nan*),' one man told me. 'If you didn't have a relative who was an official, or know someone at the factory who would put your name forward, or if you weren't taking over your father's or mother's job, then you didn't have a chance.'

In 1978, Sun Fuji, now 31 years old, was still working at the Radiant Porcelain Factory. He had switched from glazing to operating the jigger-jolly machine at the front of the line. Jigger-jolly operators had one of the jobs that locals referred to as 'bitter-dirty-tiring' (*ku zang lei*), but were paid better than other workers. Sun Fuji needed a well-paying position because he had to feed and clothe a family of four on ration coupons for one. In 1968, when Jingdezhen's Cultural Revolution violence was peaking, Sun's parents had begged their son to go to the countryside. Sun went, but he hadn't liked it and came back to Jingdezhen after only a few months – with a new wife. The Suns had two children in quick succession (1970 and 1971). Mrs Sun had a rural residence permit, which meant that she could not legally work in the city and was not entitled to ration coupons. The children's residence permits were rural too. So Sun Fuji and his wife did sideline private commerce. She sold eggs boiled in tea and hand-made slippers and whatever else she could think of. If the police caught her, they confiscated her goods and told her to go home. Sun Fuji sold the porcelain that the Radiant Porcelain Factory had given him for bonuses, and what he lifted from the factory's storage area. When he had one or two pots to sell, Sun sat outside a hotel and looked for someone whom he was sure wasn't from Jingdezhen. When he saw an outsider, he'd offer him or her the wares. 'I had customers from all over,' Sun said. 'Even Beijing. There was nowhere in Jingdezhen to buy a piece of porcelain. After I had been doing it for a while, visitors would come looking for me and ask for ceramics. I would help them out, find something at the factory to sell them.'

In 1979, Sun was finally able to switch his wife's, son's, and daughter's residence permits to urban status. Mrs Sun could then get a job in a small collective porcelain factory. On 31 December 1980, when the city government received orders to allow private enterprise, she could also sell her tea-boiled eggs without fear of consequences. Jingdezhen's Office of Statistics recorded 400 people working in private commerce in 1980. This was a tiny fraction of the city's total working population of 191,400, but still a 400 per cent increase from the previous year. By 1985, the number had grown to 6,800 (Table 4).

In May 1981, Jingdezhen's Communist Party called a meeting of all the ceramic work units administered by the city government, including china stone mines and enterprises in the surrounding villages. The Party announced that every porcelain enterprise must work to make Jingdezhen a more competitive 'brand'. The industry, leaders decided, would concentrate on producing high-quality dinnerware sets, and emphasize exports to Europe and the United States.[38] Jingdezhen's direction for the next decade was set.

## Plan and market together: The early years

When China's top leaders approved the nation's sixth Five Year Plan in 1982, government officials managed every aspect of the porcelain industry's functioning, including production levels, ware types, prices, profits and losses, taxes, hiring, salaries, and future plans.[39] The Porcelain Industry Company oversaw 52 enterprises, 23 state and 29 collective, producing everything needed to manufacture porcelain: china stone, machines and kilns, pigments, transfers, brushes, glazes, finished porcelain, and packaging. The porcelain factories alone employed 60,000 workers in 1984.[40] The city had multiple ceramics research centres staffed by scientists, engineers, artists, designers, and technicians. There were middle and high schools that began training students for the ceramics industry, ceramics vocational schools for high school graduates, and institutions offering advanced degrees in porcelain art, design, and engineering.[41] National

**Table 4** Jingdezhen employment by sector, 1978–1993[42]

| Year | Total population | Urban population | Private employees | Total workers at work units (*danwei*) | SOE* workers | Collective workers and other | Employed population |
|---|---|---|---|---|---|---|---|
| 1978 | | 524,222 | | *180,400* | *141,100* | *39,300* | 180,400 |
| 1980 | | 537,241 | 400 | *191,400* | *147,600* | *43,800* | 191,800 |
| 1985 | | 569,672 | 6,800 | *243,600* | *169,300* | *74,300* | 250,400 |
| 1986 | | 580,694 | 11,700 | | | | |
| *1990* | | | | *254,700* | *184,500* | *72,900* | |
| **1991** | **1,328,800** | **444,400** | | | ***190,800*** | ***72,000*** | |
| **1992** | **1,338,600** | **449,000** | | | ***190,100*** | ***74,600*** | |
| **1993** | **1,357,800** | **371,700** | **18,400** | ***254,900*** | ***184,000*** | ***70,900*** | 273,300 |

* State-owned enterprise.

leaders, including the General Secretary of the Communist Party Hu Yaobang, signalled the industry's importance by visiting in November 1982 and September 1984.[43]

In 1983, Jingdezhen was China's top earner of foreign currency among the nation's ceramics manufacturing sites.[44] In 1985, the value of the city's export porcelain was $20,050,000 USD.[45] Twenty-nine provinces, cities, and special economic zones sold Jingdezhen porcelain, and it was exported to more than 100 international locations in 1986.[46] Between 1984 and 1989, Jingdezhen dinnerware regularly won awards and prizes: 29 from the provincial government, 12 from the Ministry of Light Industry, eight from other branches of the central government, and four from international competitions.[47]

Wang Chunman started working at the Flourish Jingdezhen Porcelain Factory (*Jingxing Cichang*) in 1981. She was 16 years old, and had completed one month of high school. Wang was the youngest of seven children. All of her siblings were employed, and one of her older sisters who worked in a state porcelain factory sometimes gave her a little money. Wang thought it felt good to have cash in her pocket (*you qian shufu de hen*), so she asked her older brother to help her get a job. He had been a soldier and was working at the Water and Electricity Bureau. Wang's brother scolded her for wanting to quit high school. He himself had not been given the chance to attend. But Wang Chunman insisted that working was more important. Ultimately her brother asked his friend, who was the head of the Labour Bureau, to find a job for his baby sister.[48]

The Flourish Jingdezhen Porcelain Factory was a state enterprise that produced export-quality dinnerware. In September 1981, its managers submitted for official approval a list of individuals whom they wanted to hire. The head of the Labour Bureau told them that he would grant their request if they found a job for Wang, so the factory offered her a job as a gilder with large collective employee status (*da jiti bianzhi*). Wang's salary would be the same as the workers with state enterprise status (*guoying bianzhi*), but she would receive a lower retirement pension.

Gilding was desirable work. It was easier than pasting decals, the other job typically given females in porcelain factories. Gilders did not sit on the assembly line, but worked at desks in a separate section. Unlike the line workers, they could take a break or use the toilet whenever they wanted. Flourish Jingdezhen had 40 to 50 gilders, one for every two workers who applied decals. The gilders completed in three hours the number of pots that it took the decal decorators five hours to finish.

Wang started as an apprentice earning 15 yuan per month. She gave 10 yuan of that to her father, to cover her living expenses (Wang's brothers didn't have to turn over part of their salaries to her father, because they were expected to be saving for their weddings). Wang punched the clock and for her first year did whatever her master told her to do, little of which related to gilding. Sixteen young people began as apprentices with Wang, and they did not have much to do. One afternoon they decided to sneak out to the movies. When the factory manager turned up looking for them, the apprentices scattered and ran out of the theatre. The manager caught and disciplined one group, conducting a formal investigation and holding a factory-wide criticism meeting. Wang fled with a girl whose family had a personal connection to the manager. He told them not to let it happen again, and let them go.

Wang could participate in production during the second year of her apprenticeship, and by her third year she was a regular employee, and could have taken on apprentices herself. If a worker trained an apprentice, she was allowed to count the apprentice's wares as part of her own quota once the apprentice could produce pots of adequate quality, and receive a bonus for over-quota production. But without apprentices, Wang could leave work when she finished gilding her allotment of wares, except when the factory leaders decided that everyone had to work by the clock. 'Training apprentices wasn't worth it,' Wang Chunman told me. Wang earned three yuan per day, and could earn 1.8 yuan for an overtime shift. But if she took piece-rate jobs at private workshops after work, she earned even more.

A few Jingdezhen residents began opening private ceramics enterprises in 1983 and 1984, and many more followed after China's 13th Party Congress in 1987 (Table 5). Many of the people who went into private ceramics during the 1980s could not get good jobs in state enterprises: they had bad class labels, or were migrants from the countryside or workers who had been fired for having 'over-plan' births (a second or a third child). Most of the new private entrepreneurs had little capital, so they could not afford large machinery or bulk supplies. Many

**Table 5** Jingdezhen private ceramics producers, 1986–1993[49]

| Year | No. of households in private ceramic manufacturing |
|---|---|
| 1986 | 42 |
| 1987 | 60+ |
| 1988 | 152+ |
| 1990 | 726 |
| 1993 | 1,000+ |

turned to small-scale production of decorative porcelain, working as owner-operators, relying on family for labour and cash.

The private porcelain sector that emerged in the 1980s was segmented, much like the commercial industry had been prior to the socialist reforms of the 1950s. A thrower or hand-presser or someone who knew how to slip-cast would produce forms. Another person would buy and decorate them. The decorator purchased glaze from a glaze-maker, who bought materials from entrepreneurs like Wang Chunman's older sister, who opened a tiny ceramics supply store after she was fired from the Red Star Porcelain Factory in 1985 for giving birth to her second child. To fire, a private entrepreneur had to know a factory employee who worked in the kiln section. For a fee, the worker would fire privately produced wares in the factory's kilns on a Sunday, when the state and collective enterprises were closed. Jingdezhen had no private kilns until 1992, when Liu Yuanchang, a sculptor and administrator at the Sculpture Factory, bought the first propane kiln from Japan.[50] After that, private kilns became a new form of enterprise. As was true at the commercial kilns which had operated through the early 1950s, a kiln owner rented space to private ceramists. Some potters bought their own kilns and ran small comprehensive potting workshops.

Most of the private entrepreneurs made replicas of historic porcelain (*fangguci*). By his own account, the first person in Jingdezhen to produce replicas of historic porcelain was Luo Xuezheng. Luo was from Ji'an County in southern Jiangxi, a rural area with its own tradition of ceramics (black stoneware tea bowls decorated with leaves or paper cuttings). Luo attended the Jingdezhen Ceramics Institute and graduated from the art department in 1967, just as the Cultural Revolution broke out. He never told me what he did during the Cultural Revolution. In 1978, Luo was assigned a job researching historic ceramics at the Light Industry Bureau Ceramics Research Center, and began writing a book about Chinese ceramics for his employer. The Center sent Luo to Beijing to speak with curators and other specialists at the Palace Museum in 1983. He met a woman in her early forties who was running an export business. She told Luo that foreigners were eager to buy replicas of imperial porcelain. Luo realized that the Light Industry Bureau Ceramics Research Center could manufacture reproductions of Yuan, Ming, and Qing porcelain to sell. He pitched the idea to his bosses, but they refused. 'They were scared,' Luo said. 'Everyone was. The Cultural Revolution had only recently ended. Everyone had absorbed Mao's ideas and thought in terms of the planned economy. No one knew where the reforms were going.'

Luo thought the historic replica idea was too good to pass up, so he talked four of his colleagues into opening a private workshop. They studied archaeological evidence, hired ceramists who had potted and painted before mechanization, and made pots with handicraft technologies such as stick-spun potter's wheels. The workshop opened for business in 1984. 'We took it easy during the day, while we were at the research center,' Luo remembered. 'We devoted ourselves to working at night and on Sundays when we were at the workshop, figuring out how to make high-quality antique reproductions.' Luo sold pots to Hong Kong and Macao (then not part of the People's Republic of China), and the market was excellent.

In 1985, Luo and his colleagues were told either to quit their jobs at the Light Industry Bureau Ceramics Research Center or give up the workshop. The men had 'iron rice bowls', lifetime state sector employment with a pension, medical benefits, and housing, and none of them were willing to give that up for private enterprise. They sold their workshop to one of their customers, a 22-year-old rural migrant named Xiang Yuanhua.

Xiang Yuanhua was the seventh of eight children from a poor farming family. He had been unable to get a job in a porcelain factory, so had taken to buying and selling antique porcelain as a private entrepreneur.[51] When Luo offered Xiang the chance to buy his historic replica workshop, Xiang seized the opportunity. Luo maintained relations with the enterprise, using his contacts to help make the Hua Hong Porcelain Factory successful. Government officials regularly patronized the workshop and brought their guests. In 2005, many residents pointed out Xiang Yuanhua to me as one of the first people in Jingdezhen to get rich from private ceramics.

## Policy changes and competitor industries, 1984–1993

In 1984, the central government began a decade-long policy of giving municipal authorities more control over expenditures and state enterprises within broad guidelines.[52] Local officials in many cities founded public–private manufacturing ventures. The central government stipulated that the profits from such firms should be reinvested or used for public works.[53] Officials in Chaozhou and Foshan (Guangdong Province), Dehua (Fujian Province), Zibo (Shandong Province), Liling (Hunan Province), and Chongqing (Sichuan Province) established ceramics factories and gave them tax abatements and other assistance.[54] For example, the Chaozhou municipal government partnered with managers and workers at state and collective enterprises to found several new porcelain factories. Cadres purchased technology, built kilns, and supported research and development.[55] The resulting factories sold art porcelain, dinnerware, tiles, sanitary porcelain, garden ceramics, and porcelain for electronics. In Foshan, the mayor set up public–private plants, gave grants to new private businesses, and borrowed money to modernize factories.[56] The new Foshan enterprises produced architectural tiles and sanitary porcelain, both good sellers as the People's Republic had entered a construction boom. Zibo officials focused on high-tech porcelain for electronics, aerospace, and defence industries. They founded 70 technology centres to support research and development of new materials.[57]

The central government instructed Jingdezhen's Porcelain Industry Company to give the porcelain enterprises more autonomy to make decisions about wages, bonuses, and overtime, and to hire workers on contracts rather than through the permanent employment (*bianzhi*) system, on 18 May 1984.[58] Officials were directed to reduce taxes and the fees that the Porcelain Industry Company charged for distribution, so that factories could retain more of their earnings. In 1985, the central government informed the city that future funding would come in the form of bank credit.[59] The nationalized distribution network that the Party had begun in 1952 was disbanded.[60]

Municipal and provincial officials were not enthusiastic about the policies. Taxes from the porcelain industry were a major source of revenue. Already the state enterprises hired ceramists at state, large collective, and small collective worker statuses; if the factories also hired contract labourers with different wages and bonuses, the employment situation would become 'chaotic'. In 1985, cadres permitted some state enterprises and municipal villages to set up new collectives and hire 1,700 workers.[61] Official documents reported that these were poorly managed and had big losses. In 1986, a branch factory of the For the People Porcelain Factory was allowed to hire 350 workers in 'bitter-dirty-tiring' positions as contract labourers. However, the factory treated these workers as if they had permanent employment status: no one was ever fired or terminated.[62] When the ultra-modern China Wind Porcelain Factory opened in 1987, the Labor Bureau allowed the factory to hire 677 contract workers, but again the contract employees were treated as if they had state enterprise employment status.[63] In October 1987, officials experimented with allowing the Art Porcelain Factory more control over its revenues.[64] They divided the factory into 80 units and gave each unit a quota. Units that produced over quota could sell the wares on the market. Official records do not indicate whether this initiative was successful.

The central government continued to reform state and collective enterprises in the late 1980s, developing a 'managerial contract responsibility system' to give factory managers more decision-making power and incentives.[65] Ceramics factory managers in Chaozhou, Chongqing, Yulin, and other places took advantage of the policy to recruit foreign investors, mostly from Hong Kong.[66] In Jingdezhen, officials implemented the reform more slowly. In July 1991, a Shanghai-based joint venture company was allowed to rent a section of a state porcelain factory to produce sanitary porcelain.[67] In August, the China Wind Porcelain Factory, which had been in production for only four years, rented 25 per cent of its facilities to a Hong Kong company.[68] In 1992, several other state enterprises implemented what officials called 'one factory, two systems', mimicking the central government's phrase to describe Hong Kong's governance after the British colony reverted to Chinese rule in 1997 ('one country, two systems').[69] The For the People Porcelain Factory, the People's Porcelain Factory, the Ceramic Decorative Materials Factory, and the China Pottery Porcelain Factory (*Jingtao Cichang*) designated branches within their facilities that could take on foreign investment.

Jingdezhen officials did not aggressively pursue foreign investment or start public–private manufacturing ventures, but they did provide extensive credit for expanding and improving the state and collective factories. In 1989, cadres decided to turn a branch of the For the People Porcelain Factory into a new state enterprise under the management of the Light Industry Bureau. The new Jingdezhen Porcelain Factory was equipped with three production lines that used German clay dust presses and other imported technology to produce Western-style dinnerware.[70] The factory began manufacture in September 1991, and immediately won a government prize and signed a contract with Mikasa.[71] Between 1990 and 1993, city officials authorized several other state enterprises to purchase foreign technology.[72] These included Japanese technology for the China Electric Porcelain Factory, German technology for the Plaster of Paris Mould Factory, and a German kiln and French technology for the China Porcelain Capital Sanitary Factory.

Between 1987 and 1991, officials switched 12 of the porcelain factories to piped coal gas.[73] Two factories imported gas tunnel kilns from abroad. The new kilns did not require saggars. Despite this warning sign of impending change, officials mechanized, centralized, and improved the production of saggars in 1989.[74] Five years later, cadres switched 25 more tunnel kilns to coal gas.[75] In 1991, the Light Industry Ministry, Provincial Government, and relevant city units approved investment in six new projects proposed by the Porcelain Industry Company, including a china stone mine, a kiln furniture factory, a mould factory, a new production line for the Radiant Porcelain Factory, and new facilities for the New China Porcelain Factory.[76] In 1993, officials set up a research team to study how Jingdezhen could increase its sales to domestic consumers.[77]

The city government implemented several infrastructure projects from the mid-1980s through the early 1990s which benefitted the porcelain industry.[78] Municipal officials built or rebuilt 45 roads, paved many streets with asphalt or concrete, and installed 2,550 street lights. They invested in improving the city's electrical supply and telecommunications technology. They built an additional railroad line to Anhui Province. Officials applied for permission to build a new airport, and began construction on 30 September 1992. With the central and provincial governments, they spent money to improve the Chang River for shipping.

## Private porcelain manufacturing expands

Private porcelain manufacturing expanded quickly during the late 1980s and early 1990s. Some of the first entrepreneurs to open private workshops did very well. Their purchases of motorcycles and cars, trips to Macao for gambling, and home renovations made private enterprise look appealing to workers in the state and collective sector. The Communist Party's 1987 affirmation that private enterprise was a necessary part of the economy gave market activity more legitimacy. In the countryside, private agriculture was becoming less lucrative, so some farmers began mining and preparing china stone for private ceramics workshops.[79] Others migrated to the city looking for opportunities in private enterprise. While many opened small food stalls or sold clothing, shoes, or accessories, some went into private ceramics production.

Private porcelain manufacturing began with decorative wares, which were well-suited to production in small-scale workshops and did not compete with Jingdezhen's state and collective factories. Many of the ceramists at the Art Porcelain Factory and Sculpture Factory got into private production. Most took second jobs at private porcelain workshops; at the Art Porcelain Factory, 80 per cent of the employees had second jobs in 1986.[80] Both factories suffered as a result. Worker productivity dropped. Supplies of clay, pigments, brushes, and other materials disappeared, as employees stole the collective's supplies for use in private firms. Some workers gathered factory discards and sold them on the market, and a few private entrepreneurs used the names of collective factories to do business.[81] Consumers complained that Jingdezhen's porcelain industry was producing poor-quality wares.

While many private producers made traditional decorative ceramics, a new category of porcelain art soon appeared for sale: 'famous art by famous artists' (*mingjia mingzuo*). Jingdezhen's 'famous artists' included professors from the Jingdezhen Ceramics Institute, artists from the Light Industry Bureau Ceramics Research Center, and ceramists from the Sculpture Factory. These ceramists had received official recognition from the Chinese state for their skills. Several had been masters to the teenagers who began studying handicraft production in 1977. In 1986, a group of famous artists participated in a commercial exhibit in Hong Kong, and the works sold well. More Hong Kong exhibitions followed, and then shows in Beijing and Japan.[82] Some of these artists began selling their works privately out of their homes.

By 1993, the Jingdezhen government recorded 18,400 people working in private enterprise, and some private enterprise owners reported annual earnings of more than 10,000 yuan per year.[83] By contrast, the average annual salary for a state-owned enterprise worker in Jingdezhen was 2,670 yuan, and for a city-owned enterprise worker, 1,770 yuan.[84] Several neighbourhoods had become known as places to buy private ceramics: areas near the train and bus stations (which made shopping easy for brokers from out of town), or by state and large collective porcelain factories, where most private entrepreneurs fired their wares.[85]

Liu Changshan retired from his factory job in 1987 and ran a private workshop from his home. Liu had three sons, and wanted to earn money to pay for their weddings, which included ritual meals, wedding photography and clothing, monetary gifts for the brides (bridewealth), housing, and the newlyweds' home furnishings. During the 1980s and 1990s, weddings were becoming increasingly elaborate and expensive. Expectations for provisioning new couples, most of which were born by the groom's family, had increased dramatically from the simple, economical arrangements that typified China when Mao Zedong was alive. Liu decorated slip-cast statues of traditional deities and figures from Chinese opera to sell. He never needed to open a storefront or advertise his wares; his customers came directly to his house.

In November 1992, the central government's Light Industry Ministry designated Jingdezhen's porcelain industry as the most important ceramics manufacturing site in the nation.[86] To anyone who might have been concerned about the direction of central government policy, the announcement looked like a national commitment to Jingdezhen. Seventy-one per cent of all ceramists worked at porcelain manufacturing sites run by the government and part of the planned economy.[87] While Jingdezhen's private sector had grown rapidly, 90 per cent of Jingdezhen's economy remained in the state and collective sector.[88] An official work team tasked with researching the economic history of Jingdezhen's porcelain industry reported that Jingdezhen was producing more quantity and better quality every year, had higher labour productivity, and earned more from sales.[89]

The town's state and collective enterprises were different than they had been in 1973. To improve their facilities or expand, they got bank loans instead of direct government funding. Distribution was no longer the exclusive provenance of the Porcelain Industry Company: the porcelain factories managed some of their own sales. Factory managers had more say over how revenues were used. Some enterprises had contract employees, and workers paid by the piece, as well as ceramists with permanent job status. Yet most ceramists saw these changes as negligible. The vast majority of workers still received hourly wages.[90] They had guaranteed pensions, medical benefits, factory-run schools for their children – all the elements of an 'iron rice bowl'. What was new was the opportunity to earn extra cash in the private sector. To most if not all of Jingdezhen's porcelain workers, the future looked increasingly bright.

# 6

# Porcelain Capital No More, 1994–2010

The Build the Nation Porcelain Factory, Jingdezhen's first state enterprise, was one of the first to shutter in 1995. The East Wind Porcelain Factory, For the People Porcelain Factory, and the Saggar Factory also closed in 1995.[1] The following year, the Flourish Jingdezhen Porcelain Factory and the People's Porcelain Factory went bankrupt. Most of Jingdezhen's other state and collective enterprises, including porcelain factories, ceramics machine factories, the factories that made boxes and packing material for the porcelain industry, and so on, shut down in 1997. The Porcelain Brush Factory stayed open until early 1998. The Radiant Porcelain Factory was the last state-owned porcelain factory to close, in October 1998 (see Plate 14).

According to a high-ranking official in the Porcelain Industry Company, municipal officials had forced Radiant to close. The enterprise had 2,937 workers, excluding its administration. As recently as 1996, the government had described the Radiant Porcelain Factory as the 'backbone' of the local economy.[2] In 1994, officials had approved a 60 million yuan loan to Radiant for technological reforms, and both the provincial and city governments awarded prizes to the factory for their implementation. In 1995, Radiant produced eight million pieces of high-quality rice-grain-pattern porcelain, worth $1 million USD. From the time Radiant was built in 1961 to the time it closed in 1998, the factory won more than 60 national and international prizes for its porcelain. Yet the factory's annual operating costs of about 50 million yuan and its debt were too high for Radiant to stay open.[3]

The proximate cause of the closures was the central government's 1994 decision to reform banking and restructure loans. Banks were told to tighten credit to state and collective enterprises and the result was bankruptcies and mass layoffs across China.[4] In 1995, China's state sector employed 109.5 million workers; by 2002 that number had dropped to 69.2 million.[5] The state-sector manufacturing workforce went from 44 million to 15.5 million. In Jingdezhen, between 60,000 and 100,000 workers lost their jobs (Table 6).[6]

**Table 6** Jingdezhen employment by sector, 1993–2002[7]

| Year | Total population | Urban population | Private employees | Total workers at work units | SOE workers | Collective and other workers | Employed population | Daily use porcelain produced (pieces) |
|---|---|---|---|---|---|---|---|---|
| 1993 | 1,357,800 | 371,700 | 18,400 | **254,900** | **184,000** | **70,900** | | **418,860,000** |
| 1994 | 1,365,900 | 464,300 | 43,200 | **254,300** | **188,700** | **65,600** | | |
| 1995 | | | | *251,400* | *183,500* | *67,500* | | *343,400,000* |
| 1996 | 1,393,800 | | | 251,100 | | | 712,300 | 361,390,000 |
| 1997 | 1,404,900 | | | | | | 752,300 | 369,000,000 |
| 1998 | 1,423,000 | | | *173,500* | *126,400* | *47,100* | 749,300 | *372,990,000* |
| 1999 | 1,422,300 | | | *168,300* | *124,600* | *43,700* | 777,800 | *410,030,000* |
| 2000 | 1,437,800 | | | 159,500 | | | 780,900 | 457,000,000 |
| 2001 | 1,456,100 | | | 152,700 | | | 783,300 | 471,460,000 |
| 2002 | 1,506,400 | | | 138,800 | | | 785,000 | 480,000,000 |

The closures in Jingdezhen took place over three years, but most people recalled the bankruptcies as sudden and shocking. In part this was because Jingdezhen residents, like other Chinese citizens, believed that state enterprise employee status (*guoying bianzhi*) meant guaranteed lifetime employment. The idea that state-run factories could shutter was unthinkable. By 1997, the year that most of the state and collective porcelain factories closed, residents had already seen six state enterprises go under. Still, most people were unprepared. When a correspondent from *China Business News* investigated the bankruptcies more than a decade later, a number of laid-off porcelain workers told him that the state and collective sector 'died almost overnight'.[8]

Wu Yan was a ceramist at a state enterprise who was not taken by surprise. When the closures began in 1995, she was a designer at the People's Porcelain Factory, where she created forms and surface decoration. Wu was born in 1962. Both of her parents were government officials: her father worked at the Porcelain Industry Company, and her mother worked in the Tax Bureau. When, in 1977, city leaders told high-level cadres to designate one of their children to study handicrafts, Wu's parents picked her. She joined 36 young people who apprenticed to masters at the Light Industry Bureau Ceramics Research Center. Wu spent three years learning to construct forms and paint in underglaze and overglaze pigments, and then the Light Industry Bureau sent her to a ceramics vocational school where she studied art. Shortly after she graduated, the research centre decided it could not retain all 36 of the cadres' children, and Wu Yan was sent to work at the Art Porcelain Factory. She painted with overglaze enamels for the factory, specializing in famille rose (*fencai*). However Wu Yan's father worried about her exposure to lead from the enamels, and wanted his daughter to paint with underglaze blue instead. He arranged for her to change her work unit and become an employee at the People's Porcelain Factory, which specialized in blue and white. Wu Yan began as a painter at People's, and then switched to design. In 1986 she married a man who was also part of the group of cadres' children trained by traditional masters. They had a daughter in 1987.

In the early 1990s, Wu Yan and her husband worked for private entrepreneurs at night and on Sundays. As artists trained in traditional handicrafts, they had skills that were in demand. Wu Yan's husband was the one who suggested that they open their own porcelain workshop. He had been painting replicas of historic blue and white wares, and saw how lucrative the market was. Wu Yan might not have agreed if it hadn't been for the Build the Nation, East Wind, For the People, and Saggar Factories closing. She and her husband had relatives in government who had a sense of what might lie ahead for the state and collective sector. Early in 1996, the couple bought a potter's wheel and two kilns, and Wu Yan quit her job. The People's Porcelain Factory went bankrupt shortly afterwards.

Wu Yan and her husband hired throwers on short-term contracts at piece rates, and Wu's husband painted. Wu Yan was in charge of sales. She got pregnant shortly after she quit her state enterprise job: one advantage of working in the private sector was that she could not be fired for violating the One Child Policy. She paid a fine for the over-plan birth, and that was that. By 1997, Wu Yan and her husband had replicas of Yuan and Ming dynasty blue and white porcelain ready for sale. She rented a storefront near the train station. Wu Yan recalled:

> At that time, you could really make a bundle from historic reproductions. Not too many people were making replicas, and producing successful pieces was difficult. Most other private entrepreneurs were still firing in coal kilns, and many of the pots that came out were unsellable. But we had propane kilns, and our success rate was much higher. We could sell a single piece for several thousand yuan. In 1998 and 1999, we made a lot of money selling replicas to buyers from Fujian Province. Of course, those brokers really made the big money. They resold our replicas as genuine antiques, and made tens of thousands of yuan on a single piece, or even more.

Many residents in the early twenty-first century thought that Jingdezhen had returned to what the town was prior to the Communist Party's rule. Liu Zhimin, a potter in his sixties who ran a small private workshop, put it this way:

> My biggest experience in Jingdezhen has been going in a circle (*zou quanzi*). Jingdezhen in the 1950s had lots of small private workshops scattered throughout the town. Then the Party came and collectivized and changed everything, building the large nationalized factories. Then came the reforms and the factories broke up (*jieti*). Now everyone is in private business, just like before the Party arrived. The only difference between Jingdezhen today and Jingdezhen in the past is that ceramists are less skilled than they were in the 1950s. Otherwise, everything is exactly the same.

The myriad, scattered, small-scale, private workshops in which ceramists primarily used handicraft methods to produce decorative wares did resemble Jingdezhen in the past. Yet there were major differences between Jingdezhen in 2010 and what the porcelain industry had been at the turn of the twentieth century. Many more ceramists could make a living from porcelain in 2010 than was the case in 1940, when manufacturing had shrunk to levels not seen for centuries. The household ceramics that outsold Jingdezhen porcelain in 2010 were not cheap foreign products; they were made by domestic competitors in

Guangdong, Fujian, Hunan, and other provinces. The most striking difference was political. In previous centuries, the state invested in Jingdezhen's porcelain manufacturing as long as the regime had functional sovereignty. Over successive governments, officials pulled back only when the state was failing. After new leaders established new governments, the state again supervised ceramics production and spent resources on Jingdezhen. By contrast, the governing regime was firmly in control when the state pulled away from porcelain manufacturing during the mid-1990s.

## Decollectivization

In November 1993, China's Communist Party issued a blueprint for turning the state industrial sector into market-based corporations, a decision that required strong political will, as it entailed allowing more than 100,000 state enterprises to fail if they could not privatize.[9] Yet according to Western observers, the People's Republic had to privatize industry if it hoped to maintain economic growth. High-level leaders disapproved of the decision, but ultimately the Party moved forward with the plan, hoping that the private sector, which was growing at a national average of 10 per cent annually, could absorb the hundreds of thousands of workers who would be laid off. China's ninth Five Year Plan (1996–2000) announced that China's state sector should become 'modern enterprises'.[10] The central government left local officials in charge of this transformation.[11]

While thousands of state enterprises went bankrupt across China between 1994 and 2005, laying off tens of millions of workers, cadres in some places managed to transition their locales to market-driven manufacturing.[12] In Guangdong, Fujian, Sichuan, Guangxi, and elsewhere, municipal officials helped state-run ceramics factories find foreign investors, transform into shareholding companies, acquire new technologies, and produce high-demand goods such as architectural tiles, sanitary porcelain, and industrial ceramics.[13] With the help of municipal officials, private ceramics manufacturing grew rapidly in Chaozhou and Foshan, in Guangdong Province. By the early 2000s, the two cities dominated China's domestic and international ceramic sales. Chaozhou became China's largest producer of sanitary porcelain and bathroom fittings, accounting for more than half of the nation's total output and earning $1.5 billion USD in 2003.[14] In 2004, central government officials awarded the title of 'porcelain capital' to Chaozhou in recognition of the size, volume, and value of its porcelain manufacturing, stripping Jingdezhen of the designation. Over the next several years, Chaozhou's porcelain production value continued to rise, reaching $7.88 billion USD in 2011. In Foshan, manufacturers focused on ceramic tiles.[15] During the 1990s, Foshan entrepreneurs regularly sold every tile they produced, regardless of quality, because domestic demand was so strong. By 2010, Foshan manufactured 60 per cent of China's tiles. The value of its exports to the European Union alone was $2.29 billion USD. Successful market transitions like these made China a major player in global ceramics manufacturing.[16] China became the 'world's factory' in other areas over the same period, overtaking Japan as the world's second-largest manufacturer in 2006, and the United States as the world's largest manufacturer in 2010.

Jingdezhen's officials wanted the state and collective porcelain enterprises to succeed as market-based firms. Cadres borrowed heavily to update old facilities, purchase new technology, and build new plants in the late 1980s and early 1990s because they thought such investment would help Jingdezhen be competitive as an international porcelain manufacturer. When officials received word of the central government's privatization plan in November 1993, they instructed the porcelain factories to assess their assets and search for external investors.[17] In 1994, the East Wind Porcelain Factory and the Saggar Factory began building storefronts and workshops on their sites to rent to private entrepreneurs.[18] The Jingdezhen Porcelain Factory and the Sculpture Factory attracted local businessmen who rented sections of the factories to manufacture porcelain for private sale.[19] Yet the only state enterprise to locate significant foreign investment was the China Wind Porcelain Factory, which had entered production in 1985 as a manufacturer of dinnerware sets, souvenirs, and decorative porcelain. On 1 July 1993, a Hong Kong businessman reopened the plant as the China Peak Ceramics Company, manufacturing architectural tiles.[20]

What Jingdezhen produced – and did not produce – contributed to the demise of the state and collective sector. In 1979, at the beginning of the Reform Era, the central government instructed city officials to concentrate on daily-use porcelain for export, and let go of architectural, sanitary, and industrial porcelain production.[21] In 1986, cadres decided that Jingdezhen would primarily manufacture mid- and high-quality dinnerware sets.[22] Yet in England, Europe and the United States, companies that made fine porcelain dinnerware were downsizing, merging with other firms, and going bankrupt.[23] The elaborate tableware that dominated porcelain sales in the 1950s was eclipsed in the 1980s by more-casual ceramic 'lines'. By the time that Jingdezhen officials realized architectural tiles, sanitary porcelain, and ceramics for industry and electronics were the porcelain commodities in demand, central government officials had already cut off credit and dumped the state sector onto the market. Only 2 per cent of what Jingdezhen produced in the mid-1990s was sanitary or architectural porcelain, and 3 per cent electrical ceramics.[24]

As factory after factory shuttered, municipal leaders tried desperately to convince the public that Jingdezhen's future was positive. In March 1997, Jingdezhen's Deputy Mayor described the closures as 'changing outdated management policies to help stimulate the local economy'.[25] 'While rejuvenating the ceramic industry,' he continued, 'the city has selected helicopters, minivans, refrigerator compressors, building materials, and pressed salted chicken as Jingdezhen's pillar industries.' He added that the city would also develop ceramic tourism.

To the tens of thousands of porcelain workers who lost their jobs, such rhetoric was unconvincing. I began interviewing laid-off workers in 2003, five years after the last state and collective porcelain factories had closed. Many blamed local officials for the collapse of large-scale porcelain manufacturing. While some criticized cadres for their ignorance about markets, or their failure to update ware types, many believed that official corruption had caused the bankruptcies. A number of residents told me that cadres and managers sold off the factories' equipment and facilities and pocketed the cash. Some former workers said that officials called in their relatives to remove wares from factory storage and sell them privately. Several ceramists pointed out that while ordinary workers struggled to find jobs after the layoffs, many factory administrators founded new private enterprises in the former state and collective factories.

Hong Ling was a woman in her late thirties who had been laid off from the large collective Golden Wind Porcelain Factory when I met her in 2005. She minded a storefront for her husband's younger sister, who ran a small ceramics factory that produced soup sets. Hong Ling and two of her older brothers had worked in state and collective porcelain factories, and all were laid off within an 18-month period. The oldest brother had been a state enterprise worker at Build the Nation, one of the first factories to go bankrupt. I asked Hong what her brother had done when he lost his job. She said that immediately after the factory's closure he went into private business with the former factory head and two other high-level cadres. The four ran a porcelain enterprise on Build the Nation's premises. A decade after Build the Nation had shuttered, her brother still made 'good money', Hong told me, pointing out that he had just purchased a home in an expensive new development near one of Jingdezhen's fanciest hotels.

## Porcelain workers during deindustrialization

Jingdezhen had about 700 private porcelain entrepreneurs at the time of the layoffs.[26] Most produced decorative wares, sculpture, and historic reproductions.[27] These businessmen hired handicraft potters and painters, but had no use for workers who operated the machines that produced dinnerware. Relatively few private entrepreneurs manufactured tableware in 1995 and 1996, as the state and collective factories were producing ample quantities of inexpensive china. The private dinnerware factories that existed, and the new ones that came into being as laid-off workers scrambled to find ways to support themselves and their families, employed no more than 100 ceramists at most. Many had only a dozen workers. The bosses hired ceramists by the order, and paid piece-rate wages. The jobs were temporary and lacked benefits or insurance.

Gender and age strongly affected whether or not workers from the daily use porcelain factories could find employment. In general, the ceramists who did the worst were older men whose careers had been in mechanized mass-production. Sun Fuji was one of many men who could not find work after he was laid off. He was 51 years old when the Radiant Porcelain Factory closed, laying off Sun, his wife, son, daughter, and son-in-law. During Sun's final few years as a state-sector employee, he worked in the clay-making section. His main task was hauling clay from the production area to the front of the assembly line. Sun had requested to switch to this job, because it meant he could earn full wages without putting in a lot of hours. As a clay-hauler Sun worked two or three hours daily, and then he could leave the factory. He spent his free time repairing his house, selling bits of porcelain, and walking around town. After Sun was laid off, he approached the private entrepreneurs who had decided to rent portions of the Radiant Porcelain Factory and asked for a job. One entrepreneur made blue-and-white rice-grain-pattern rice bowls, another made soup spoons, and a third made clay to sell. None would give him employment. 'They all said I was too old,' Sun told me. 'They said that there were plenty of younger men whom they could hire for heavy work. Why would they want someone in his fifties?' Sun tried several other private bosses who were starting small-scale dinnerware production businesses in the former work units, but no one would hire him. After 35 years in porcelain manufacturing, Sun was unemployable. He was also too young to draw his pension. Male workers were not eligible for retirement until they reached the age of 60.

I met many men who shared Sun's predicament, men who were in their late thirties or older who lacked handicraft skills. These men accepted odd jobs wherever they could, and they switched work frequently in an unending quest for a stable income. Some of the younger men purchased motorcycles and operated them as (illegal) motorcycle taxis (*modi*). A motorcycle taxi driver charged one or two yuan for a ride on his bike across the city. Some men hunted for antique pot sherds at construction sites and along the river. When they had collected a decent quantity of sherds, they laid them out on a cloth on the street and peddled them to passers-by. Other men sold paintbrushes, second-hand textbooks, or other odds and ends. Some men opened mom-and-pop food stalls or convenience stands with their wives. Some simply lived off their wives, taking over the food shopping and cooking. Many laid-off porcelain workers applied to their neighbourhood committees for low-income support (*di bao*).

Some younger laid-off workers, in their thirties and early forties, decided to open their own porcelain businesses. Others participated in retraining programmes. The Labour Bureau provided a limited number of retraining opportunities, restricted to laid-off workers who had higher educational credentials and could learn fields like computers or electronics. Most of the retraining programmes were private and taught participants porcelain painting. In 2005, one local official told me that more than 10,000 people were studying overglaze enamel painting in Jingdezhen.

Wang Guo'an and his wife worked at the same state porcelain factory and were laid off simultaneously. Wang was 43 when he lost his job. Immediately after he was laid off, Wang applied for jobs at private porcelain factories and was told that he was too old. Wang's wife was a few years younger than him. She initially tried to find work operating a cash register for a store, and was told that no one would hire a women older than 35 for that job. After months of fruitless searching, the Wangs decided that they already had some skills as ceramists, and were too old to retrain. Wang Guo'an knew the basic procedure for slip-casting. Mrs Wang could apply transfers and decals. They decided to open their own workshop, producing low-quality slip-cast jars and urns simply decorated with underglaze blue transfers for bulk sales to retailers. The couple worked long hours, and did not make a lot of money, but told me that they were getting by. 'We don't go hungry,' Mrs Wang informed me. 'If you are willing to work hard in porcelain, you will make enough to eat' (*fen fan chi*). The Wangs' son was in high school and his parents hoped that he would pass the college entrance exams, go to college, and move permanently away from Jingdezhen. They did not want him to pursue a career in porcelain.

Many of the people who went into small-scale porcelain manufacturing in the years following the bankruptcies did not make it as entrepreneurs. I conducted much of my fieldwork serially, making multiple short visits to Jingdezhen over a seven-year period. Each time I returned to Jingdezhen, I discovered small ceramic businesses that had failed. Many times I was unable to find the porcelain entrepreneurs whose workshops I had visited only a year before. The sheer volume of people who shared similar skills, and opened ceramics businesses at the same time, made ceramics manufacturing extremely competitive.[28] Laid-off ceramics workers who made porcelain dinnerware in the former state and collective factories, or who used slip-casting to make low-quality art or tableware in small workshops, participated in a 'race to the bottom', trying to undersell competitors by making coarse porcelain for sale at lower and lower prices.

The female factory workers who were laid off had better luck in the private labour market than their male counterparts. Women older than 35 had a harder time finding jobs than younger women, but if they were willing to stick transfers or decals on pots in private factories, or to work as cleaners, then they could find work (see Plate 15). Women who pasted transfers or decals worked long hours. They were expected to cut the transfers in the evening when they were at home, to prepare for application the following day. Cleaning jobs were somewhat available, as only the more highly ranked hotels insisted that their maids be young women. Some women found work looking after storefronts, typically for their relatives. Minding a storefront was considered quite good employment, relative to transfers or cleaning.

After the Golden Wind Porcelain Factory closed in 1997, Hong Ling initially went to work for her brother, pasting decals in his private enterprise. Hong found working for her brother harder and more tiring than her job in the collective factory had been. She worked longer hours, and her brother expected that she and the other women would apply 1,000 decals per day. Hong could only manage 300. After a while, Hong asked her brother if she could switch to working half days.

> I couldn't take the work, it was just too hard (*wo chi bu liao nei ge ku*). For someone like me, who was in her 40s, and accustomed to the pace of work in the collective factories, the job was too much. I said to my brother that I needed some free time (*kongxian*), and I wanted to work half days. He wouldn't let me! Can you believe it?! My own brother wouldn't let me work half days. He said that all the other workers wanted free time too, and if they saw me getting favourable treatment they would complain.

Hong quit working for her brother. Her husband had gone into porcelain manufacturing with his younger brothers, and he was making enough money that she could stay home. 'I had a great life,' Hong Ling told me. 'I wouldn't get out of bed until 10 or 11 a.m., just in time to fix lunch for my husband. When he went back to work, I went out to gamble at mahjong. I'd come home to fix dinner, and then head back out to play mahjong again. Now that was comfortable!' After six years, her husband put his foot down, complaining that not only did he have to support Hong, he had to support her mahjong addiction as well. Hong could easily lose 100 or 200 yuan in a single game. She and her husband had a big fight, and Hong ended up minding the storefront for her husband's younger sister. Her duties were not very onerous. She passed the time knitting or chatting with the other women who watched porcelain stores on the same street, and from time to time someone came by for a pickup. Hong got a small salary, which she kept to spend as she chose. In 2005, she'd switched her leisure activity to ballroom dancing, attending a nightly dance held in a room above a department store. The monthly dance pass was a lot cheaper than mahjong.

Hong explained that when she began working for her husband's sister's business, her sister-in-law received a three-year tax abatement. The city government had a policy that former state and collective workers were entitled to three tax-free years after they found new employment. Hong and her husband's sister took Hong's certificate showing that she was a laid-off collective sector worker to the city government, and then the sister got the tax abatement. Hong then added that she did not pay rent on her house, and had never paid rent during the six years that she, her husband, and their son had lived there. Hong Ling lived in public housing and the rent was supposed to be 15 yuan per month. 'When someone comes by to collect the rent, I tell them that I am a laid-off worker and they leave me alone,' Hong said. 'Lots of laid-off workers don't pay rent. Doesn't matter if they live in public housing or housing owned by the porcelain factories. They just refuse to pay.'

The most successful laid-off porcelain workers were ceramists who had artistic or handicraft skills. The artists and artisans who were laid off from the Sculpture Factory and Art Porcelain Factory transitioned into private enterprise without much difficulty. At the Sculpture Factory, for example, many former workers were sculptors, mould-makers, hand-pressers, and painters in overglaze and underglaze enamels. When the factory closed, the laid-off workers received a small payout, to terminate their employment status as large collective workers. The former workers received a one-time opportunity to buy a finite amount of medical insurance and a pension. Ceramists who were close to retirement were excluded; the factory allowed these workers to keep their large collective employee status, and they would receive their regular pensions for the years they had worked. Those who had retired before bankruptcy were unaffected by the factory's dissolution. Factory leaders then divided the premises into smaller units and rented them to private individuals. When I visited the Sculpture Factory between 2003 and 2005, many of the private ceramists who produced and sold wares there were former employees. The bulk of porcelain sold at the Sculpture Factory was 'new style' art and 'famous art by famous artists'. Some artists from other parts of China, such as Beijing, Nanjing, and Shenyang, also rented workshops in the Sculpture Factory. In 2005, a Hong Kong ceramist opened a space in the Sculpture Factory for artistic residencies (I discuss this more in the following chapter).

One well-known success story in the former Sculpture Factory was sculptor Liu Yuanchang.[29] Liu was born in the city of Ji'an, Jiangxi Province, in 1939, and admitted as one of the Jingdezhen Ceramics Institute's first sculpture students in 1959. Liu became an employee at the Sculpture Factory in 1964. He spent his entire career there, receiving numerous awards for his work between 1964 and 1996, when the Chinese government awarded him the title of national-level master artist (*guojiaji gongyi meishu dashi*). When the Sculpture Factory closed in 1997, Liu was the factory head. He returned to full-time sculpting, renting a small space on the premises to serve as his studio. In his first years as a private entrepreneur, Liu produced rather traditional works that were strongly influenced by the sculptures he had made for his former work unit. After a few years, he adopted a more abstract style.

In 2003, local cadres asked Liu Yuanchang to rent a larger space in the Sculpture Factory, and he agreed. Liu made both collectible works of art (*zuopin*) and a production line (*chanpin*). He employed 20 ceramists to make moulds, hand-press clay forms, produce glazes, paint, and fire his wares. Liu concentrated on sculpting forms, and on occasion, made moulds to cast them. As his daughter Liu Danyuan explained to me, Liu could produce wares more quickly if he hired others to complete various steps under his instructions, and Jingdezhen was full of ceramists looking for that kind of work. Liu sold his art directly from his studio. His star-studded career, and the fact that he was the only sculptor among China's 12 national-level master artists, meant that his works commanded high prices.

In 2004, Liu expanded his workshop to offer a limited number of residencies for visiting artists. He prepared some simple rooms, and charged guests a fee for a stay of one to two months. Several foreign potters and sculptors came to work in his studio, and Liu regularly hosted visiting graduate students and artists from other parts of China as well. He earned an excellent living from his sculpture, but continued to live in cheap public housing so he could devote his resources to the workshop. Liu's daughter told me that her father hosted foreign ceramists, Chinese artists, and graduate students not for financial reasons, but because he wanted to promote (*xuanchan*) Jingdezhen and its porcelain.

## Repurposing former SOEs and collectives

Immediately after decollectivization, the fate of Jingdezhen's shuttered state and collective factories was uncertain. Initially, some laid-off workers believed that the enterprises would reopen. Ceramists with state and large collective employee status had been guaranteed lifetime employment and benefits, and retained some stake in their work units' futures as long as they kept their employee status, even if the work unit itself was closed. Many of the factories rented sections to private entrepreneurs who ran small manufacturing firms, typically producing a single ware type or small lines (such as tea sets). Some private businessmen produced under the names of the factories whose facilities they rented. For example, the Red Radiance Porcelain Factory had been a large collective. After it closed its doors, a private individual contracted to use its facilities. He called his factory Red Radiance, and made some of the same wares that the former large collective had made. Red Radiance had some dinnerware sets that were popular in China, and the entrepreneur was busy filling orders. He hired workers on short-term contracts and paid them piece-rate wages. In 2000, the Chang River flooded, and part of the factory was submerged. The entrepreneur ceased production.

After a few years, cadres began selling parts of the former state and collective factories, and in some cases, entire sites. One of the first to be sold was the Flourish Jingdezhen Porcelain Factory, where Wang Chunman had worked. In 2000, the Bureau of Industry and Commerce purchased the site for ten million yuan. The factory was demolished and the site turned into a fruit and vegetable market. According to Wang Chunman, the money for the sale went to the former factory leaders, and the workers never saw a penny. In reality, much of the sale price must have gone to pay Flourish Jingdezhen's debts.

The East Wind Porcelain Factory was sold in 2002. A private developer, in collaboration with city officials, purchased a portion of East Wind's facilities, demolished them, and built a set of storefronts that he called the International Market. The former state enterprise had a favourable location near Jingdezhen's main street, the People's Square, and the train station. By 2003, many of the new storefronts had been rented, and most sold porcelain (see Plate 16). The International Market also contained a few restaurants, a retraining school, and a branch of the Bank of China.

When the East Wind Porcelain Factory was sold, cadres used a large portion of the money to pay the SOE's outstanding debts to the Bank of China and other creditors. Then factory leaders made one-time payouts to the former workers who were near retirement age, in lieu of pensions. For younger workers, cadres purchased short-term medical insurance for five to ten years, effectively buying out the former workers' rights to pensions and terminating their relationship with the factory. These laid-off workers gave up their state enterprise employment status and assumed responsibility for their own futures. Cadres used the remaining funds to buy a small plot in the suburbs and set up a porcelain factory. The new factory operated under a new name, was a stock company rather than a state enterprise, and employed about 100 ceramists. It produced on a much smaller scale than East Wind had.

Other state and collective enterprises followed a similar process when they were sold. The Art Porcelain Factory, the New China Porcelain Factory, the Flourish Jingdezhen Porcelain Factory, and the Ceramics Machine Factory all set up new, small-scale factories in the suburbs, with about 100 employees, after they had paid their debts and (in some cases) made payouts to former workers. According to a reporter for *China Business News*, the ten state porcelain factories alone spent more than 70 billion yuan on medical insurance, pension payoffs, and other expenses.[30] Many laid-off workers complained about the small size of their payouts and the time-restricted medical insurance that they received. Yet ceramists who had worked at state and collective factories that remained unsold were eager for such sales to occur. Until the enterprise facilities sold, laid-off workers received no compensation, regardless of whether or not the factory's facilities were rented to private entrepreneurs. In 2010, many state and collective porcelain enterprises remained standing, including the Build the Nation Porcelain Factory, the People's Porcelain Factory, the Red Flag Porcelain Factory, the For the People Porcelain Factory, and the Radiant Porcelain Factory.

## Private enterprise: Historic replicas

In the mid-1990s, most private porcelain producers made art porcelain in small-scale workshops.[31] Many more entrepreneurs made replicas of historic porcelain than new-style ceramics or 'famous art by famous artists'. According to Fang Lili, who did field research in Fanjiajing, a neighbourhood (once a village) dominated by replica makers, the annual income for proprietors of workshops making mid-quality antique replicas was between 20,000 and 60,000 yuan in 1996 and 1997.[32] A thrower working for hire could earn between 120 and 1,200 yuan per day, depending on the size and quantity of the wares he threw.[33] A painter earned between 1,000 and 10,000 yuan per month, depending on how skilled s/he was, and how quickly s/he worked. These incomes equalled or exceeded the average annual salary for a state or collective enterprise worker in Jingdezhen: in 1993, a state enterprise worker had earned 32,040 yuan, and a large collective worker, 21,240 yuan.[34]

Laid-off workers flooded the art porcelain field, dramatically inflating the numbers of small workshops and the number of ceramists who worked for hire. In 1993, the entire city had 400 private ceramics enterprises; by 2008 that number had risen to more than 2,700.[35] In 2010, news media reported more than 4,000 porcelain workshops in Jingdezhen.[36] Workshop proprietors in Fanjiajing reported that they earned half as much in 1996 as they had in 1995, and 1997 brought further drops in earnings.[37] In-migration of youth from the countryside, many of whom had only completed junior middle school, exacerbated Jingdezhen's overproduction of art porcelain. While these rural migrants lacked experience in ceramics, many were willing to take any job, work any hours, and accept any wage, so that they could learn how to make porcelain and eventually open their own small enterprises.

When I first visited Jingdezhen in 2003, hundreds of workshops sold underglaze blue and polychrome overglaze replicas of Qing porcelain, particularly from the Kangxi, Yongzheng, and Qianlong periods.[38] Wares ranged in quality from cheaply made slip-cast replicas to high-quality hand-thrown reproductions. Ninety-eight per cent of production took place in small, family-owned workshops that had fewer than five permanent employees. Proprietors were owner-operators, employing their skills in at least one portion of the production process (often potting or painting, or if the business were a commercial kiln, firing). Bosses tacked up signs outside of their storefronts advertising work for specific skills: an underglaze blue painter experienced in painting dragons, a hand-presser, a mould-maker, a calligrapher to inscribe reign dates on the feet of historic replicas. An entrepreneur also used his or her networks to locate employees. Ceramists who did not own their own businesses were temporary workers, paid by the piece and employed by the order. They had no guarantees that a workshop boss would rehire them, although some proprietors insisted that if they were satisfied with a ceramist's work, they were morally obligated to employ that individual when they had another order that s/he could fill. Still, I witnessed loud public arguments when a proprietor did not rehire a ceramist whom s/he had previously employed, evidence that a sense of moral obligation did not always outrate profit motive.

In Chinese, the name for replicas of historic porcelain is *fangguci*, 'porcelain copied from antiquity'. Ceramists who worked in historic replica workshops had to have good skills as copyists. In 2005, I apprenticed to a 24-year-old man who founded a small business making replicas of Yuan dynasty porcelain. Yang Yajun had begun working in porcelain at the age of 16, when he apprenticed as a hand-presser in his uncle's (mother's sister's husband's) antique replica workshop. He was a motivated apprentice and learned hand-pressing in one year. Yang told me that after he had learned how to do hand-pressing, he would watch the other ceramists employed by his uncle, and stay late at the workshop practising their skills. His uncle closed his workshop in 2004, and after a few months of freelancing, Yang opened his own place in May 2005. As an owner-operator, Yang formed pots by hand-pressing, made his own clay, models, and moulds, hand carved surface decoration, and fired pots – although he occasionally hired a kiln master to manage firing (see Plate 17).

Yang Yajun made high-quality replicas, and he used photographs to help him copy forms. Typically a photograph of a ware in a museum, an art dealer's catalogue, or an online site would list at least one dimension of the pot. Yang would use that dimension to figure out the relationship between the size of the image and the size of the piece, and then determine the other dimensions. He modelled the piece in clay, using his experience to decide how much bigger he needed to make his model to account for shrinkage during firing. Then Yang used plaster of Paris to make a mould.

Yang Yajun also copied the defects on Yuan pots. He studied pot sherds and photographs so that he knew which defects occurred during the Yuan dynasty and which did not. For example, he sometimes made his underglaze blue pigment a little greyish, because Yuan pigments sometimes came out greyish rather than blue, and the use of the grey made his pots more convincing. Yang also aged his pots, using his own recipe to give his wares a finish that was softer and less glassy than a contemporary piece. He was fully aware that the dealers who purchased his replicas might choose to sell them as genuine antiques. 'It is none of my business what they do with my pots,' Yang told me. 'I sell them as replicas.'

In 2005, Yang Yajun's wife was studying underglaze blue painting but could not paint well enough to produce high-end pieces. When Yang had a load of greenware that he wanted decorated, he hired underglaze blue painters. Some were painters who had worked for his uncle. One, Wang Zhuwen, was a woman in her late twenties whom Yang's master had introduced to him. When a painter worked on a replica, s/he would first photocopy a photograph that showed the surface decoration of an antique, resizing it to be the exact dimensions of the original pot. The painter then used pouncing to transfer the image to the surface of a piece of greenware, pricking holes in a translucent paper along the lines of the image, laying the pierced paper on the surface of the pot, and using a sponge to apply an ink wash over the holes. When s/he removed the paper, an ink outline of the surface decoration appeared on the surface of the greenware. A painter painted over the ink with cobalt pigment; the ink burned off during firing.

Certain surface motifs were very common on Yuan pots, such as cloud forms, waves, lotus borders, and vine scrolls. Underglaze blue painters who specialized in Yuan dynasty wares had practised these motifs so many times that they painted them freehand. During my research in Jingdezhen I saw apprentices sitting at their desks day after day, repeating cloud forms on a flat block of greenware. When the block was full, the apprentice scraped the surface clean so s/he could begin practising again.

When Yang Yajun made replicas for sophisticated customers, he made pots that adhered to the Yuan-dynasty blue and white genre but did not exactly duplicate any existing piece. This was especially crucial for the surface decoration. Yang explained that the freehand, simple style of Yuan painting on ceramics was much harder to copy than the formal patterns found on Qing pots. Yuan painters worked within conventions, but they painted without pouncing, and their brushstrokes showed vigour and individuality. A Yuan-dynasty replica painter had to copy, but s/he also had to inject the painting with spirit. If s/he copied the surface decoration faithfully, but the brushstrokes lacked vitality, the painting would look 'dead' (*siban*) and the reproduction would be lower quality.

Not all private producers of historic replicas hand-potted, but the majority did paint by hand, unless they were making very cheap replicas indeed. Aging replicas was also very common. I saw replica makers using citric acid, dye, traditional Chinese medicine, shoe polish, and tea to age wares. Others beat their pots with something hard (such as a piece of kiln furniture), or rubbed wares with sandpaper. Some ceramists adhered substances to their replicas' surfaces to make them appear to have been 'buried for centuries', as one proprietor put it. Since most genuine antiques were excavated, some porcelain brokers wanted pots that looked dirty. On one occasion, I watched a producer use an electric sander to shape a sherd from a piece of Qing porcelain, and then glue it into the foot of one of his wares. If a buyer took a core sample from the foot of this pot, the test results would come back that it was a genuine Qing dynasty porcelain.

Ceramists were sanguine about these activities. If a buyer wanted wares to look old, then any entrepreneur who wanted to sell wares would age them.

## Porcelain fakes on the art market

Jingdezhen's historic replicas were usually purchased by brokers and resold. Some dealers travelled north to Beijing's famous antique market (Panjiayuan), others went south to Guangzhou (Canton) and Hong Kong, and others headed for other cities. It was an open secret that the best antique replicas were resold as antiques. I heard several small workshop owners say that what happened after they sold a pot was none of their affair. As long as their buyers knew what they were getting, the entrepreneur had done nothing wrong. I met ceramists who were perfectly willing to serve clients who brought in antiques for copying. The potter was paid for making a copy, and did not ask what the customer planned to do with the replica.

I heard many stories about ceramists who worked actively to transform their replicas into antiques, although only one potter admitted to doing this himself. Several ceramists named high-end replica producers who also worked as porcelain appraisers. These men (all were male) were said to use their connections to auction houses and museums to sell replicas as antiques. For example, a ceramist might know an auctioneer who was willing to sell a replica as an antique in exchange for an extra commission. Many ceramists told me that Chinese auction houses did not scrutinize the wares they sold, and that they were willing to cut deals with replica makers to sell pots as antiques. A ceramist might also give some of his replicas to an appraiser, who, for a fee, claimed that the pots were part of his private collection of antiques. The appraiser-collector then donated part of the collection to an art museum, and sold the rest through an auction house. Among the wares sold were some replicas, now part of a 'museum quality' collection of antiques. Part of the proceeds went back to the original potter, and the rest were kept by the appraiser-dealer. Locals also told me that replica-makers took advantage of overseas relatives to add value to pots. An overseas relative would claim that a replica was a family heirloom, giving the pot a false provenance. Others talked about making deals with museum curators. One painter suggested to me that I should use my museum connections in the United States to make money selling 'antique' Jingdezhen porcelain (I declined).

Some Jingdezhen copies found their way into auction houses, dealerships, and museums, where they were shown as genuine antiques. Many of these replicas went undetected, but a few splashy cases made international headlines. In July 2013, for example, authorities in Hebei Province closed the Jibaozhai Museum, which had forty thousand fake objects in the collection, and only 80 authentic.[39] On 22 May 2014, *The Guardian* reported that Chinese police shut down the Lucheng Museum in Liaoning province because almost one-third of the 8,000 objects on display were fake. Both of these museums contained porcelain from Jingdezhen.

As the market for antique Chinese porcelain heated up, so did the business of making high-quality replicas. After Jingdezhen ceramists got word that Christie's had sold a fourteenth-century Yuan dynasty jar for $27.7 million USD in July 2005, those who could replicate Yuan porcelain turned immediately to reproducing the ware.[40] When the taxi driver turned art collector Liu Yiqian purchased a Chenghua period chicken cup for $36.05 million USD at auction in Hong Kong, he broke the record for the highest price reached by Chinese porcelain. Liu planned to house the cup in his private museum in Shanghai.[41] But he also commissioned replicas from Jingdezhen to sell. By April 2015, Liu had amassed nearly $800,000 USD from the sales of his chicken cup replicas.[42]

## New-style artist entrepreneurs

As the historic replica field grew overcrowded, competitive, and less lucrative in the late 1990s and early 2000s, ceramists searched for new markets. Between January and October of 2003, 1,500 entrepreneurs from Jingdezhen went abroad to sell porcelain.[43] Numbers increased the following year, according to local reports. These ceramists paid 10,000 yuan to purchase licenses from the government for international sales. Most went to destinations in South and Southeast Asia. Government documents and Jingdezhen residents both stated that there were 'problems' with the earliest ceramists who went abroad. Several ceramists told me that the first entrepreneurs who travelled internationally had been 'uncivilized' and 'uneducated' (*bu wenming, meiyou wenhua*), and left a bad impression on their foreign hosts. In addition, none of the sellers had wanted to bring wares home, so as their departure time neared, they discounted their prices. This in turn made sales difficult for the next group of entrepreneurs. As a result, officials increased their supervision of international ceramics sellers.

Some ceramists responded to pressure in the historic replica field by switching to new-style art porcelain. In the mid-1990s, Jingdezhen had three ceramists who were recognized at the national level as artists and 18 who had been recognized at the provincial level.[44] The artist Tai Cilu, president of the Jingdezhen Ceramics Institute, sold one of his works for $60,000 USD, which was the highest price that a piece of contemporary art porcelain had ever commanded on the international market. Some artists from the group of cadres' children trained in handicraft methods turned to 'famous art' porcelain shortly thereafter. In 1999 these ceramists made less than Wu Yan and her husband did selling historic replicas, but by 2003 'famous artists' commanded higher prices per piece. Wu Yan decided to let her historic reproduction business go and retrain herself as a painter of 'new style' (*xinshi*) art porcelain.

When Wu Yan entered the new-style art porcelain field, the government held examinations to designate ceramists as local, provincial, or national-level artists. Wu Yan took the examination in 2004 and received the title of provincial-level ceramic artist. When I met her in August 2005, she was still retraining herself and reconnecting her networks so that she could succeed as a 'famous artist'. She ran a high-end porcelain shop where she sold her own and others' art, including historic replicas left over from the workshop she had run with her husband. Wu Yan worried that she had entered the 'famous art' field too late. Some of her friends who sold famous art in 2003 and 2004 made more money on a single work, and sold more work, than the market could bear in 2005. Still, between 2005 and 2010, Wu Yan developed a clientele. Her art included large porcelain tiles painted with branches of cherry tree blossoms in underglaze blue and red (see Plate 18), urns decorated with peonies and butterflies in vivid overglaze enamels, and a range of other wares. When I visited Wu Yan in the winter of 2010, her business had been profitable enough that she could afford to buy a second place, a brand-new two-bedroom apartment in a high-rise minutes away from her combined studio and storefront.

As 'famous art by famous artists' gained popularity, and the prices garnered by titled artists and professors at the Jingdezhen Ceramics Institute and the Light Industry Bureau Ceramics Research Center rose, artists at the Light Industry Bureau Ceramics Research Center were instructed to create at least one new work each month.[45] The central government cut funding, telling departments to pay for themselves, and the Center decided that artists' salaries should be based on sales of their works. The institute looked for other sources of revenue too, producing ceramic cooking pots and leasing out facilities to private entrepreneurs. Research dropped precipitously, and leaders did not replace equipment that became outdated.

Unlike historic replicas, new-style art was supposed to be 'created' (*chuangzao*), not copied from antiquity. But in the first decade of the twenty-first century, many new-style ceramic artists copied one another's works.[46] When some new-style artists visited galleries or attended art shows, they photographed the work of other ceramists for 'reference' (*cankao*), taking note of which wares sold and which did not. Some purchased catalogues of new-style porcelain art and created works based on what they saw, at times with limited modification from the original. When I first visited the Light Industry Bureau Ceramics Research Center's art gallery, I was shocked to see that the surface designs by its 'famous artists' were more or less the same as the art porcelain that I had seen for sale all over Jingdezhen. When I visited the Jingdezhen Ceramics Institute's galleries, particularly in the student gallery, I saw a broader repertoire of art porcelain, perhaps because commercial success was not a degree requirement. In fact, many Jingdezhen Ceramics Institute students marketed their wares at a Saturday market, and some made tens of thousands of yuan annually from selling porcelain art.[47]

In the first decade of the twenty-first century, many ceramists felt strongly that it was difficult to be successful as an artist entrepreneur.[48] The porcelain market was intensely competitive and immoral. If a ceramist managed to produce a work that sold well, she or he could not prevent others from copying the work and taking away from the original artist's market share. Ceramists with a commitment to artistic creation did their best to create works that were individual, but admitted that their efforts were sure to be defeated by other ceramists. Some artists hoped to limit copying and counterfeiting of their wares by closing their work spaces to everyone except relatives and close friends. I heard about one ceramic painter who changed his signature every three months to foil copyists reproducing his painted porcelain tiles.

## Investing in Jingdezhen

In the first decade of the 2000s, Jingdezhen officials began to bring in the foreign investment that had eluded the city during the 1990s. By 2008, the city had attracted foreign investors from Taiwan, the United States, Japan, and Thailand.[49] Among these the largest was Franz Collection Incorporated. Taiwanese artist and entrepreneur Francis (Li-Heng) Chen founded the porcelain company in 2002, locating its design studio in Taiwan, headquarters in San Francisco, and production facility in Jingdezhen.[50] Franz manufactured porcelain tableware, vases, decorative accents, and jewellery. In the first decade of the twenty-first century, luxury department stores such as Harrods, Saks Fifth Avenue, Bergdorf

Goodman, and Bloomingdales, fine porcelain dealers in France, Germany, Austria, and Switzerland, and numerous stores in China, Taiwan, and Japan sold Franz ceramics. Franz's Jingdezhen workforce was not large, but the company was a good citizen. For example, in 2004 Francis Chen assisted municipal officials hosting a commemoration of 1,000 years of porcelain production in Jingdezhen by 'bringing together porcelain manufacturers from all over the world at a grand, one-of-a-kind celebration'.[51] The company also planted trees as a carbon offset for firing, including thousands of Chinese olive trees, red maples, loropetalum, and taxaceae.[52]

Analysts of Jingdezhen's ceramics industry lauded its acquisition of foreign investment, but encouraged the government to do more. For example, Xiao Gongduo and Guo Guo'an of the Jiangxi Ceramics Industry Corporation called on the government to manage Jingdezhen's brand or 'trademark' ('made in Jingdezhen'), facilitate market distribution, build infrastructure, and locate funding for research and development.[53] They felt that Jingdezhen had begun to recover from the state and collective sector collapse, but needed more investment, better organization, and more advanced technology. Unless Jingdezhen could scale-up its small workshops, they warned, the city could not reclaim its 'former glories'.[54] Li Songjie, Wang Shujing, and Li Xinghua, a group of professors and senior lecturers at the Jingdezhen Ceramics Institute, also insisted that the government should do more for Jingdezhen. They called on officials to supervise porcelain manufacturing and protect copyright, restrict the production of low-end ceramics, and help small workshops develop into larger factories.[55]

The Jingdezhen residents with whom I spoke between 2003 and 2010 were almost unanimous in their desire for government officials to do more for porcelain manufacturing. Ceramists working in dinnerware factories and art porcelain workshops, teachers at ceramics education institutions, and local officials told me that the government needed to take part in porcelain production. Many thought that city officials should rebuild the large-scale porcelain enterprises. Others wanted the government to regulate ceramics, for example by enforcing copyright protections for artists, ensuring quality control of china stone and other supplies, and limiting the number of people who could produce the same ware. Over and over I heard ceramists complain that city officials did not care about (*bu guan*) porcelain manufacturers, artists, and artisans. A smaller group complained that the efforts officials did make, such as hosting an annual Ceramics Festival or trying to make all the ceramists move their kilns and production sites out of the city centre, were superficial or wrongheaded.

Li Chunlin, born 1970, painted and sold new-style art porcelain decorated with overglaze enamels in a tiny workshop in downtown Jingdezhen. For five months in 2005, I visited Li's workshop-cum-storefront two or three times a week. Li had been laid off from the Art Porcelain Factory, where he had worked in the sales department until the factory went bankrupt. Not long after Li lost his job, his wife was laid off, when the large collective factory where she worked closed after it lost its state-enterprise clients (the factory made boxes for shipping porcelain). When I met Li Chunlin, he supported his family, which included one daughter and a second on the way, with his earnings as a commercial painter and shop proprietor. One afternoon, I asked Li to share his thoughts about Jingdezhen's porcelain industry. He thought for awhile, and then replied:

> The city government is not paying enough attention to the ceramics industry. The government should protect, nurture, correct, and assist with porcelain production. It is the officials' job to take care of the industry. Instead they do nothing. They go to work, eat and drink well on public money [*chi guojia de qian*], take home their salaries, and they don't care. The government only intervenes when there is a crisis, and by then it is already too late. Officials don't think about the laid-off workers. We got next to nothing when we were laid off, and now we are left on our own. The government doesn't care if you live or die.

As the first decade of the twenty-first century concluded, officials told the media that they wanted to develop aviation, tourism, and ceramics in Jingdezhen.[56] But as one analyst pointed out, no one would think of Jingdezhen when it came to aviation. When people heard 'Jingdezhen', they thought 'porcelain'. Most local residents and analysts believed that Jingdezhen's future had to be porcelain. The question they confronted was what a Jingdezhen with a fully private porcelain manufacturing sector would look like over the long term. Jingdezhen and its residents were accustomed to state investment in and management of ceramics. If the government did not support research, direct production, or commission large orders, would the city continue to employ large numbers of ceramists or maintain a large industry? Without the state, could Jingdezhen ever again become China's porcelain capital?

# 7

# From Porcelain Capital to Heritage Site

When the People's Republic of China relinquished most forms of industry to private actors, central government officials had already begun encouraging the development of tourism. Officials first acknowledged tourism as a segment of the national economy in China's seventh Five Year Plan (1986–1990), and the central government implemented several policies to stimulate tourism during the 1990s.[1] These included working with foreign investors to build state-level resorts, reducing the length of the Chinese citizen's work-week, mandating two weeks of paid vacation for employees, and extending the duration of three public holidays.[2] As geographers Tim Oakes and Travis Klingberg discuss, Chinese leaders wanted to stimulate the growth of a leisure economy in which 'culture' was a major resource for development.

During the 2000s, the Chinese government stressed tourism's role in the nation's continued economic growth. When the 2008 global financial crisis began to affect the PRC, officials spent $60 billion USD on infrastructure development projects across the country, in the hopes that domestic tourists would plug the revenue gap caused by declining foreign purchases.[3] The following year, the State Council released its 'Guidelines to Accelerate the Development of the Tourism Industry', in which provincial governments were told to encourage the construction of hotels and convention centres, provide more transportation services, build recreation and entertainment outlets, improve tourism services, and enhance vocational training for hospitality workers.[4]

Jingdezhen officials turned to urban renewal and heritage to attract tourists, using 300 million yuan in central government funding to 'preserve the porcelain capital'. Municipal officials demolished large tracts of the city, replacing narrow alleys with wide streets and collaborating with commercial real estate developers to build new housing and business districts. Cadres developed the Ming and Qing dynasty Imperial Kiln (*Yue Yao*) as a historic attraction, and constructed a new 'Old Street' nearby. Officials commissioned public art, such as underglaze blue porcelain lampposts, to adorn the city, and invented rituals to promote Jingdezhen's imperial past.

According to a professor of ceramics history at the Jingdezhen Ceramics Institute, tourism surpassed porcelain manufacturing as Jingdezhen's top source of revenue in 2003. The numbers of tourists visiting the former porcelain capital rose from 100,000 in 2000, to 6,690,000 in 2006, to 20 million in 2012 (Table 7). Jingdezhen earned 12.52 billion yuan from tourism in 2012, more than four times as much as the 2.88 billion yuan that ceramics generated.[5] The vast majority of tourists were from other parts of China, but a noticeable group were artists from Australia, Britain, Canada, China, Denmark, Italy, Japan, Korea, Mexico, the Netherlands, New Zealand, Norway, the United States, and elsewhere. Some artists came to tour, but many wanted to experiment with Jingdezhen's famous ceramic medium.

The idea to offer porcelain capital residencies for artists came from the private sector, not the government. Jingdezhen's first artistic residencies were the brainchild of artist Li Jianshen (also known as Jackson Li), who opened the Sanbao International Ceramic Art Institute in a former village on the outskirts of Jingdezhen in 1998. By the time I visited in 2003, Sanbao had already hosted artists from Australia, China, Europe, Korea, and North America. A few years after Sanbao's genesis, sculptor Liu Yuanchang and ceramist Carolyn Cheng set up two more sites for artistic residencies, both in the former large collective Sculpture Factory. Cheng's Pottery Workshop, which opened in May 2005, added a 'Saturday Creative Market' in 2008 as a venue for artists to sell their work. The weekly event received media attention from Chinese and international newspapers, blogs, and travel websites.[6] By 2010, Jingdezhen was well established as a location for international ceramics residencies.

Chinese and foreign artists took advantage of Jingdezhen's abundant and cheap supply of raw materials and labour to create works and set up semi-permanent and permanent studios. Some local ceramists got work from these

**Table 7** Tourist visitors to Jingdezhen, 2000–2014[7]

| Year | Visitors |
|---|---:|
| 2000 | 100,000 |
| 2003 | 500,000 |
| 2006 | 6,690,000 |
| 2012 | 20,000,000 |
| 2014 | 25,850,800 |

visitors, but far fewer than had been employed by the state and collective porcelain enterprises. Some working relationships between local ceramists and visiting artists endured for years; one celebrated example is the Old Wei Workshop (*Lao Wei Tang*) which began working with conceptual artist Ai Weiwei in 2006.[8] Local ceramists created a range of objects for Ai, including one hundred million porcelain sunflower seeds for his Tate Modern exhibit in 2010. While the project employed more than 1,600 ceramists over a two-and-a-half-year period, Liu Weiwei, the proprietor of the Old Wei Workshop, could not rely exclusively on Ai Weiwei for income. He also produced 'high-end counterfeits of Ming and Qing blue-and-white for insertion into Chinese auctions'.[9]

## Officials redevelop the porcelain capital

To make the city appealing to tourists and outside investors, local cadres adopted principles that had been used in other urban areas. They decided that production sites, including small-scale porcelain manufacturing workshops, should move out of the city centre to the suburbs, and be replaced by new commercial spaces, service industries, and high-rise housing. Public art, parks, and tourist areas would make the city an attractive destination, with wider roads to facilitate private cars. In the suburbs, officials wanted developers to construct large residential areas comprised of free-standing houses with private garages, lawns, and gardens. City leaders relied on central government funding, local incentives (such as tax abatements) for private investors, and money from external developers from Hong Kong, Jiangsu, Zhejiang, Guangdong, and other parts of China to accomplish this transformation.

Between 2003 and 2010, demolition areas and new construction checkered Jingdezhen. Some new projects worked very well, such as the International Market on part of the former East Wind Porcelain Factory grounds. Officials planned to expand the renovation into the area to the north and west of the market, which included a former state enterprise printing press, porcelain workshops, small storefronts, and housing. During the second half of 2005, officials and residents were arguing over how much owners would be compensated: the city offered people with private houses in the area 50,000 yuan to leave, but residents refused, saying the sum was too little. As a number of people pointed out to me, 50,000 yuan would not be enough to buy a place in that location after the reconstruction. By January 2010, clearing was underway.

Some areas of new construction failed to attract renters or buyers, such as the new development to the west of the river that locals called 'West of the River' (*hexi*), and officials referred to as 'Pottery Town' (*taocheng*). West of the River contained storefronts and storage areas housed in quasi-traditional-style architecture and was located near the bus station. It did not include sites for production. Officials wanted the development to become a porcelain market, but the lack of manufacturing spaces made the area unattractive to many ceramists. Jingdezhen's private porcelain entrepreneurs were accustomed to selling directly from their workshops, conducting production and sales more or less simultaneously. Entrepreneurs who rented or bought in West of the River would have to transport wares from their workshops to the storefronts, or pay someone else to haul them. When West of the River was built, potters and painters clustered in neighbourhoods that were known for specific products: historic replicas, tiles, 'famous art by famous people', sculpture, large wares, and so on. By contrast, the area west of the river had been farms and had no history of porcelain manufacturing. Ceramists did not want to move to locations that lacked a reputation for porcelain. In addition, many private ceramists had their own propane kilns which were not portable. While the cost of building a propane-canister kiln had decreased dramatically since Jingdezhen first imported the technology in 1992, kilns were still a major expense. A proprietor who owned a functional kiln would not willingly abandon it, and officials refused to compensate private entrepreneurs for their kilns if they relocated from the city centre.

Municipal officials thought that the best way to attract tourists was to tear down most of Jingdezhen's existing buildings, regardless of whether or not they were of historic interest. For example, curator and Sinologist Archibald Dooley Brankston photographed Jingdezhen's elaborate opera house during his 1937 visit.[10] Officials divided the opera house into ten housing units for ten families during the 1950s, when the city's population expanded rapidly. A former resident described growing up there as 'very lively'. In 1988 or 1989, the city tore down the opera house and replaced it with a multi-storey apartment building. Another historic site that was demolished was the wood-burning Yishan kiln built during the Guangxu period of the Qing dynasty (1875–1908) and used continuously until its final firing in 1998. Some locals wanted the city to preserve the kiln, but officials destroyed it circa 2001.[11] Elsewhere in the old city, neighbourhoods that contained wood-burning kilns, old houses constructed from kiln bricks and saggars, two-storey wooden houses that had been occupied by porcelain brokers during the early twentieth century (and perhaps as far back as the Ming), large traditional houses lit by sky wells (*tianjin*) in the ceilings, temples, and the smoke stacks and production sites of the former state and collective enterprises were demolished.

City officials chose the imperial past for heritage tourism, even if officials had to build 'the past' themselves. Jingdezhen's Old Street was a good example of newly constructed 'imperial' heritage: a row of shops built in an architectural style reminiscent of elite buildings in 'traditional' China, with a *pailou*-style gate, curved eaves on the roofs, and red-painted wooden columns and railings. Old Street was not a hit: locals complained that the merchandise sold there was too expensive, and tourists rated it 'average' or 'poor' on travel websites. For example, one Trip Advisor user, writing on 22 March 2014, said that there was 'nothing to see here' and that 'all Chinese cities seem to have one of these'.

Officials focused on the Hutian Ancient Kiln Site, the Ancient Kiln Folk Customs Museum, and the Imperial Kiln as the centrepieces for tourism. They spent money to prettify these sites by adding quasi-traditional architectural elements, flagstones, and gardens. On special occasions, and during the annual Ceramics Fair, they enacted quasi-traditional rituals for the enjoyment of visitors, including worship ceremonies for the God of Wind and Fire and kiln recertification ceremonies.

Hutian's trajectory as a heritage site began on 30 November 1959, when the Jiangxi Provincial Government declared it a conservation area.[12] Jingdezhen cadres gave Hutian more attention after Deng Xiaoping voiced concern that Jingdezhen's handicraft traditions were disappearing, opening Hutian to the public not long after his 1973 visit. When I first visited Hutian in 2003, the site was simply presented and had the atmosphere of an archaeological excavation. By 2010, officials had provided Hutian with a number of new buildings, attractive landscaping, and more elaborately staged exhibits.

Officials began creating the Ancient Kiln and Folk Customs Museum in the late 1970s, salvaging portions of several old workshops and moving them to a large park west of the Chang River where they recreated a comprehensive potting workshop of the sort that Jingdezhen had during the late Qing dynasty. When the museum opened in the 1980s, more than 300 ceramists, many of whom had been working before 1949, demonstrated the use of handicraft tools at the site. The Ancient Kiln Porcelain Factory, as employees and locals called it, also employed young people, for example middle school graduates who began as apprentices in 1981.[13] From its beginnings, the museum-factory included commercial sales: the employees who demonstrated handicraft traditions produced wares to sell. During the 2000s, private entrepreneurs rented out various portions of the Ancient Kiln. These included producers of historic replicas, a wood-burning kiln master, ceramists who used the spaces for exhibitions and workshops, and porcelain sellers who marketed wares made outside of the park. The Ancient Kiln Folk Customs Museum sold blue and white porcelain, underglaze red porcelain, rice-grain-pattern porcelain, coloured glazed wares, and polychrome enamelled wares. In 2005, some entrepreneurs were selling porcelain decorated with a pattern that had been created for then-president George W. Bush. Shopkeepers told me that foreign tourists were their best customers, particularly Americans, Germans, and Japanese.

Starting in 2002, foreign tourists at the Ancient Kiln could pay a fee to try their hand at potting with traditional tools.[14] Guests could also pay to videotape employees making pots. To serve visitors who wanted to experiment further with porcelain clay, officials founded a Ceramic Art Center adjacent to the museum in 2004. The Center had modern equipment rather than handicraft technology. Staff said that many domestic tourists with private cars drove to the Ceramic Art Center on the weekends, looking for 'something fun to do' with their kids. Tour groups visited the Ceramic Arts Center on their way to or from the Ancient Kiln Folk Customs Museum. The centre hosted groups of primary and junior middle school children and children's summer camps. Few visitors ever completed their pots: the facility was full of greenware and bisques that guests had left unclaimed.

The city government decided to reframe its annual Ceramics Fair as a tourism event in the 2000s, linking it to the Ancient Kiln Folk Customs Museum. Cadres founded the fair in October 1990 as a trade opportunity for local producers, outside buyers, and ceramics companies from other places, including multinational firms.[15] Now officials added exhibitions, demonstrations, and rituals, promoting the event as a holiday. During the 2005 Ceramics Fair, officials held a ritual to the God of Wind and Fire at the Ancient Kiln. In 2013, they organized a kiln reactivation ceremony for 'three royal kilns built and used during the Ming and Qing dynasties'.[16] In other years, cadres arranged porcelain orchestra concerts, in which musicians played instruments made from blue and white porcelain.[17]

The central government authorized excavation of the Ming and Qing imperial manufactory in 1989.[18] Archaeologist Liu Xinyuan, who had investigated the Hutian Ancient Kiln Site in the 1970s, headed the team responsible for the dig, which included archaeologists from Beijing University and the Jiangxi Provincial Institute of Cultural Relics and Archaeology. National media reported the 'discovery' of Jingdezhen's Ming-dynasty imperial kiln, as well as the late Qing dynasty Jiangxi Porcelain Company (a short-lived effort by officials and businessmen to revive the imperial manufactory) in 2003.[19]

The imperial manufactory's 'discovery' was not news to people living in Jingdezhen. Locals said that everyone knew the imperial manufactory was underneath the city government offices. Prior to the official excavation, private entrepreneurs dug at the site, looking for antiques to sell and sherds to use as references for historic replicas. The area was riddled with tunnels from private excavations, so many that portions of the site collapsed, and some people were injured. 'It's only because people got hurt that the city government finally decided to excavate it,' commented one ceramist. In fact, official plans had more to do with promoting tourism.

On 31 December 2003, city officials held a small ceremony to commemorate the find. Several government officials, members of the archaeological team, professors from the Jingdezhen Ceramics Institute, a group of foreign artists in residence at the Sanbao International Ceramic Art Institute and I all attended. Several officials and archaeologists gave brief speeches emphasizing the site's importance and praising the government for undertaking the dig. At that time, the primary area open to visitors was the Dragon Pearl Pagoda. The pagoda exhibited photographs of the excavation and pots that had been reconstructed from the site's waste area, where imperial porcelain workers smashed porcelain deemed of insufficient quality for the emperor. Signage and facilities for visitors were limited, particularly for those who could not read Chinese. When I revisited the site in September 2004, visitors could walk directly up to the excavation pit, which was covered by a simple plastic roof. By the following year, officials had built showier structures, mounted more signage, installed large bronze statues depicting imperial officials near the site's entry, and lit the entrance with colourful lights (see Plate 19).

Public art played an important role in the city's redevelopment efforts. Officials oversaw a number of art installations, including a 'ceramic culture wall' at the Chang River Plaza, which dramatized each stage of porcelain production in 'ancient times', as the Deputy Party Secretary of the Cultural Affairs Bureau put it. Teachers and students from the Jingdezhen Ceramics Institute were commissioned to design and mount the piece. Another installation was a '1,000 years of kiln fire' porcelain mural that lined a new road in the western part of the city. Officials also mounted bronze statues depicting handicraft artisans completing different stages of pre-mechanical porcelain manufacturing in the People's Square, and commissioned sculptures made from porcelain dinnerware which they installed in the People's Square (see Plate 20). Other public art included a large sculpture of a Taiping kiln, abstract porcelain art, and porcelain lampposts and stoplights.

## Public opinion about Jingdezhen's redevelopment

Ceramics entrepreneurs, laid-off porcelain workers, rural migrants, academics, and officials expressed a range of views about Jingdezhen's reconstruction. Some residents described the city as cleaner and less chaotic, and highlighted improvements to the transportation system: a highway between Jingdezhen and Wuhan, a highway to Nanjing that was under construction in 2005, daily flights between Jingdezhen and Shanghai. These improvements facilitated porcelain commerce. Some locals commented that Jingdezhen's public art communicated the city's distinctiveness. As one recent rural migrant put it, 'Those porcelain lampposts give outsiders a good impression of Jingdezhen. They help visitors recognize that Jingdezhen is special.' The many photographs of Jingdezhen's porcelain lampposts on websites like Pinterest and Trip Advisor, and the frequency with which Jingdezhen travellers mentioned the lampposts in blogs, attested to his views.

Others thought the renewal projects had erased Jingdezhen's special characteristics. 'So much of the development makes Jingdezhen look like any other city,' one ceramist sighed. 'Jingdezhen used to have a link with the past, with its tiny alleys, the old buildings made from kiln bricks, all the different kinds of houses. It's a pity that's all gone. Now officials just play at making things look old. It's a kind of game.' An administrator at the Jingdezhen Ceramics Institute agreed. 'Our municipal officials have very little interest in preserving the past. But Jingdezhen's history is what makes us special. The old buildings were what made us different from other towns. Now we have a lot of empty new developments, and no atmosphere. We have a problem with the quality (*suzhi*) of our city officials. They refuse to talk to anyone who knows anything.' One retired porcelain worker complained that the city government appeared to believe that Jingdezhen was measured by the width of its roads and the number of foreigners who visited. 'Even now there is still so much heritage here,' he went on.

> Look, in this neighbourhood [Worker Farmer Village (*Gongnongcun*)] there used to be a huge temple complex, the Jingde Temple. There is still a temple building left over there. This temple is really old, even older than Jingdezhen. If the government restored it, worshippers and tourists would visit. They'd burn incense and spend money, and this area would become a destination. A 50,000

> yuan investment would be enough to get the temple started again. But city officials are short-sighted and they don't think about these things. They don't even know that a temple existed here. The land has already been sold. A developer is coming in to build apartments, and that will be the end of the temple. The people who move here will be young, and they won't know about the Jingde Temple. But it will bring them bad fortune just the same.

Some locals complained that the reconstruction did nothing to help citizens achieve a comfortable life (*xiaokang shenghuo*), the standard that Deng Xiaoping set out as a goal of the market reforms. Both ceramists and officials told me that they thought the redevelopment efforts were unconnected to practical needs. For example, when the city decided to build West of the River, no one had tried to determine whether or not the pre-existing amount of retail space was adequate, or asked the current ceramics entrepreneurs what they wanted or thought. The government simply built the new commercial area and waited for ceramists and buyers to come and use it. 'This development was more of a dream than a plan,' one official said to me. 'The new construction has exceeded the reality.' In other instances, ceramists complained that they were forced to go along with redevelopment to the detriment of their livelihoods. When a private real estate developer wanted to build on Shuguang Road, for example, city officials insisted that porcelain sellers vacate the area, which had been a thriving porcelain market.

## Private entrepreneurs and artistic residencies

Artistic residencies were a kind of tourism that affected ceramists more than other heritage activities. The idea to host ceramic artists for short residencies in Jingdezhen came from Li Jianshen. Li was a student at the Jingdezhen Ceramics Institute in the 1980s, where he took his BFA and MFA degrees. In 1991, Li met American ceramic artist Wayne Higby, a professor at the New York State College of Ceramics at Alfred University, at China's first international conference on ceramic art. Higby encouraged Li to come to Alfred University for graduate work, and in 1994 Li became the first Chinese artist to take an Alfred University MFA.[20] Many years later, Li told a *China Watch* reporter, 'My early experiences in the US really opened my eyes. At that time, art residencies were well established in the US and many other Western countries, but the concept was still new in China.'[21] Li decided to experiment with artistic residencies in Jingdezhen.

In 1993, Li purchased three of four residences in a small village near Jingdezhen called Four Family Village (*Si Jia Li*). He redesigned them into an artist's colony for Chinese and foreign guests that opened in 1998. During the first five years of operations, the institute hosted more than 500 international artists, including at least one group of students from Alfred University.[22] Sanbao International Ceramic Art Institute was located about 20 minutes away from the centre of Jingdezhen, in the direction of Hutian. Officials paved the road to the institute in 2003 to support Li's endeavour. The colony was comprised of beautiful wooden structures, set among forested hills and rice fields (see Plate 21). Its main residential building had a sunny paved courtyard intersected by a running stream with a plank bridge. During December of 2003, the courtyard was filled with a large art installation entitled 'the spirit of Jingdezhen', created from rows and rows of small deity statues, several ceramic pieces, and other objects by Walter McConnell, a professor of ceramic art at Alfred University. Inside the main building, art and the materials for producing art were everywhere: porcelain art and artefacts, brushes for painting and calligraphy, ceramic statues, and more.

Li Jianshen told me about his idea of a 'working ceramics village' on 30 December 2003, explaining that he wanted to preserve local heritage as a 'living tradition' rather than a 'museum piece'. Li designed Sanbao to give visitors a taste of the traditional Chinese scholar's life by living in a beautiful wooden house simply furnished with traditional chairs, where residents could hear the pounding of stream-powered water hammers crushing porcelain stone, and see the forest, hills, rice fields, village houses, and sheds where farmers made bricks of china stone to sell to urban ceramics entrepreneurs. He intended Sanbao to be 'nostalgic', evoking a 'way of life that people have had for 1,000 years'. Yet as my tour of the colony demonstrated, Sanbao was also meant to be comfortable for foreign guests, with flushing toilets, a shower, electric blankets, a ping-pong table, and a soda machine. It was also a place for ceramists to work, with potter's wheels, stoneware and porcelain clay, bats and banding wheels, and a large kiln. At the time, Li was building an exhibition gallery and a wood-burning dragon kiln up the hillside next to the colony, on land he had purchased with help from the local government. In subsequent years, Sanbao expanded further to include a bar and restaurant, a gas kiln donated by the Japanese company Shimpo, and four water-powered hammers for pulverizing porcelain stone.[23] Li saw himself as helping local officials envision 'an alternative way of developing Jingdezhen'. 'Modernization doesn't have to look just one way,' he said. Rather than tearing down old structures and replacing them with high rises, private developers and government officials could preserve some of Jingdezhen's special features and make the city an attractive destination.

Li exerted himself to ensure that his guests had opportunities to encounter local heritage, handicrafts, and art, in addition to the experience of living and working at the colony. For example, during my short visits in 2003 and 2004, Li took Sanbao's visitors, and me as the accompanying anthropologist, to see the Imperial Kiln, a fifteenth-century residence located near the Imperial Kiln, the Ancient Kiln Folk Customs Museum, the Hutian Ancient Kiln Site, the exhibition galleries at the Jingdezhen Ceramics Institute, the Hutian Ancient Reproductions Factory (which specialized in large porcelain vessels taller than two metres), the former Gaoling mine, the historic town of Yaoli (which included the Cheng lineage house), and a historic commercial kiln in Raonan. He organized a concert of Jingdezhen's porcelain orchestra at Sanbao, and a demonstration of handicraft methods by local ceramists.

Numerous artists who completed residencies at Sanbao wrote blogs or published accounts about their work and experiences.[24] Many mentioned the places where the resident artists they met came from, which included Australia, Canada, China, Mexico, the Netherlands, and the United States. I visited Sanbao only sporadically during the years that I did ethnography in Jingdezhen, yet even so I met ceramists from Australia, Canada, China, Finland, Hong Kong, Korea, Italy, Norway, and the United States. For example, Norwegian artist Ole Lislerud, a professor at the Oslo National Academy of the Arts who is known for his large-scale architectural installations, arrived at Sanbao on 1 January 2004, at a time I was visiting the institute. Lislerud had begun creating large porcelain plaques (or, as some have called them, 'tiles' or 'murals') in Jingdezhen two years previously, hiring locals to assist him and perform various aspects of production.[25] Local tile-makers made the ceramic 'canvas' with which Lislerud experimented as painter, graphic designer, and architect.[26] The creative possibilities facilitated by this method occupied Lislerud for more than a decade over several return trips to Jingdezhen (see Plate 22).

In 2004, the Sanbao International Ceramic Arts Institute lost its status as the sole site for private artistic residencies in Jingdezhen. That year, sculptor Liu Yuanchang began hosting visiting artists in his workshop at the Sculpture Factory. The residencies that Liu provided were also cultural experiences, but his guests focused on creating works. Like Li Jianshen, Liu helped his guests find local ceramists to complete various portions of their artistic production. As Liu's daughter Liu Danyuan put it, 'We are a bridge. We help foreign artists get what they need so they can work. We help them find raw materials, ceramists to make things for them, throwers, mould-makers, glaze-makers, firing – whatever our foreign artists need.'

In May 2005, ceramic artist Caroline Cheng opened another site for artistic residencies, the Pottery Workshop, which like Liu Yuanchang's studio was located in part of the former large collective Sculpture Factory. Cheng was already operating pottery workshops in Hong Kong and Shanghai when she asked ceramist Takeshi Yasuda to direct a Jingdezhen Pottery Workshop. I met with Yasuda and other members of the Pottery Workshop staff and toured the site in September 2005. A staff member explained why Cheng had wanted to open a Jingdezhen branch. The foreign artists based at her Shanghai location had frequently travelled to Jingdezhen. 'Jingdezhen is where all the materials come from,' he continued. 'Jingdezhen is the place that all the foreign visitors want to see.' The new facility had three main goals: to offer residencies for foreign artists, to expand the size and scope of Pottery Workshop operations, and to 'have a commercial aspect'. At the time, the workshop was filling an order of ginger jars and plates for a Hong Kong hotel.

In 2005, the Pottery Workshop offered residencies for foreign artists, an illustrated lecture series by the visitors, and frequent exchanges between the guests and students at the Jingdezhen Ceramics Institute. By 2008, the organization included additional studio space, and a Workshop Café where visiting artists could interact with 'the hand-chosen, cream of China's young ceramists', and a Saturday Creative Market heavily patronized by students from the Jingdezhen Ceramics Institute.[27]

Pottery Workshop staff explained that the international exchanges benefitted everyone involved. On the one hand, the Pottery Workshop gave Jingdezhen ceramists a 'window onto the world' and opportunities for inspiration from international visitors, particularly to students at the nearby Jingdezhen Ceramics Institute, who could engage with foreign ceramists, practise their English, and help the staff. For example, when Australian ceramist Robyn Phelan had a residency at the Pottery Workshop, two ceramics students assisted her with sourcing materials and equipment, and answered questions about ceramics and local practices.[28] On the other, visiting artists could work with 'traditional materials', meaning the clays, glazes, and pigments used in Jingdezhen, and the inexpensive yet expert local porcelain workers whose skills could help guest artists realize visions that were too difficult to materialize elsewhere. 'It is a lot easier to get things done in Jingdezhen,' a Pottery Workshop employee put it. 'It's cheaper, and you can find someone to do whatever you need. Certain kinds of projects are possible in Jingdezhen that just could not happen back home in Vancouver.'

Pottery Workshop visitors had opportunities for tourism at museums, factories, and shops, but spent most of their time on artistic production. Phelan wrote that most of the visiting artists came to Jingdezhen with specific projects in mind, and the staff at the Pottery Workshop quickly introduced them to the potters, glazers, or other ceramists who would help them achieve their goals. 'The work ethic at the residency was admirable,' she stated. 'Most are in the studio by eight a.m. and return to work in the evening [after dinner].'[29] Shannon Sullivan, a professor of art at the College of the Redwoods, explained to reporter Ken Weiderman in 2015 that during her Pottery Workshop residency she 'contracted out' parts of 'the creative process' so she could make the best use of her limited time.[30] Sullivan 'challenged herself to work with the unique qualities of Jingdezhen porcelain and embrace the division of labour that has made the porcelain city so successful', telling Weiderman, 'You can easily visit Jingdezhen as an artist or designer and think like a project manager. It's amazing how good you can get at something when you share the labour so you don't have to be really good at every little facet of your process . . . it's a very efficient way to use people power.'

By 2015, the *Global Times* reported that more than 20,000 'Jingdezhen drifters' (*Jingpiao*), including 1,200 foreign artists, regularly visited the city to 'soak up the atmosphere', 'set up their own porcelain brands', and establish studios.[31] A long list of international ceramic artists had hired Jingdezhen potters, mould-makers, glazers, painters, kiln masters, and other porcelain workers to help them create works.[32] According to visiting curator Anna Wu, 'The availability in Jingdezhen of an encyclopaedic range of process-specific skills . . . are Jingdezhen's most important attribute and hold the key to its future success.'[33]

Some visitors who experimented with outsourcing creative labour had mixed feelings about the results. In some cases, their disappointment derived from the gap between local and external models of being a ceramist. Non-university-trained ceramists in Jingdezhen had an extended period of repetitive and monotonous practice at their particular skill. For them, becoming an expert meant being able to replicate results exactly: they wanted to create wares that achieved perfection every single time, an ideal of manufacturing that the imperial manufactory had promoted for centuries. By contrast, many foreign artists preferred their work to have an element of surprise. For example, a staff member at the Pottery Workshop told me that the visiting ceramists were finding the Jingdezhen glazes 'a bit too predictable', so they were mixing them together 'to see if they could get some more interesting results'. When Phelan described working with a local glaze master at the Sculpture Factory, she wrote that he would 'only work to the point of his own experience rather than to achieve the end that I envisioned . . . If more time and works had been made available, I would have loved to see him push past the boundary of the familiar.'[34]

Other international ceramists felt that using local talent to make art compromised what it meant to be an artist. For example, Li Jianshen said to a reporter from *Time Out Shanghai*, 'Some people look down on you for making your own work. But if you like cooking, do you hire someone to cook in your kitchen? Do you pay someone to fuck for you?'[35] For this kind of artist, the ideal ceramist was the studio potter who engaged in every aspect of a work's creation from clay-making through final firing.

## Local perspectives on visiting artists

For most ceramists and workshop owners who learned porcelain outside of university, the visiting artists were more important as a source of income than for creative inspiration or 'international exchange'. The kinds of jobs that private entrepreneurs and ceramists for hire did for these artists were similar to those they did when hired by professors at the Jingdezhen Ceramics Institute. Locals were pleased to have such work, especially when such jobs turned into long-term relationships with repeated opportunities for revenue generation. Very few artisans felt that they had something to learn from either the international ceramists or the university and fine-arts institute artists. The former state and collective factory workers, rural migrants and younger ceramists who had trained by working for private entrepreneurs regarded themselves as much more skilled than their artist-employers.

Many of these ceramists found the art of foreign visitors unintelligible. Take Wang Zhuwen, the porcelain painter who made Yuan dynasty historic replicas in the workshop where I apprenticed. She visited the Sanbao International Ceramic Arts Institute to see a friend, and looked at some porcelain art by a European ceramist. 'The art was little piles of clay with a tiny bit of carving at the top and on the sides,' Wang said. 'To me, it looked like shit. I mean it literally looked like pieces of shit.' Wang went on to contrast 'Western art' with 'Chinese art'. 'Western art is really abstract,' she said. 'By contrast, Chinese art, or at least traditional Chinese art, is very direct and clear. A fish is a fish, a peony is a peony. It's all very easy to understand.' Other ceramists were intrigued by the foreigners' work. For example, Liu Zhimin was an entrepreneur in his sixties who regularly did jobs for visiting artists. 'I've been in the porcelain industry for 40 years,' Liu told me.

> I know the manufacturing process from start to finish. I was a sculptor and a mould-maker when I worked in a state enterprise porcelain factory, and I've taken every opportunity I've been offered to learn more about porcelain. I'm a little different than most of the ceramists you meet. I want to make porcelain that is special (*you tese*). I want to use my imagination, give my porcelain my own flavour (*ziji de weidao*).

Some ceramists who had artists as clients told me that the artists were very strict about protecting their work. The ceramists were forbidden to replicate the work, sell misfires or seconds, or produce similar pieces. One of Liu Zhimin's clients said that if he found any of the pieces that Liu made for him on the market, he would make Liu return all the money he had paid him in the past. The same skills that made outsourcing elements of the creative process attractive carried within them the threat of copying and counterfeiting.

## The porcelain tourism capital

Cultural tourism in Jingdezhen employed far fewer people than manufacturing had, as is typical at industrial heritage sites.[36] Some laid-off porcelain workers participated in the tourist sector by finding employment as maids in hotels, starting small private restaurants, or operating taxicabs to serve tourists. I met two former state and collective employees who tried working as tour guides, but quickly changed to other lines of work. They found that being a tour guide required education, cosmopolitanism, and better social networks than they or most laid-off workers possessed. Jingdezhen made money from tourism, but the former state and collective enterprise workers and ceramists were not the primary beneficiaries.

A small number of lucky ceramists got government commissions to make public art, which brought in cash and heightened their reputations; the ones who I met were all educated at university. A few of the better-connected private porcelain producers, such as the Jiayang Porcelain Company and the Hua Hong Porcelain Company, became regular stops for tourist groups. Local officials who knew the proprietors brought official visitors and promoted the companies' manufacturing sites as destinations to tourist agencies. The Sanbao International Ceramic Arts Institute, the Hutian Ancient Kiln Site, and the Ancient Kiln Folk Customs Museum also benefitted from the drive to build tourism, becoming the most frequent stops in Jingdezhen for companies that organized ceramic tours.[37] Sanbao was a private artists colony, but its owner cultivated excellent relations with local officials. Hutian and the Ancient Kiln were official historic sites, with opportunities for some private ceramists to sell works to visitors.

Officials and private entrepreneurs promoted imperial and handicraft heritage. No one, it seemed, considered Jingdezhen's modern industrial heritage to be worth seeing or saving. I heard a few residents jokingly refer to the smoke stacks of the former state porcelain factories as 'historic relics', but their offhand remarks were not linked to efforts to preserve or commemorate the city's industrial past. As architectural historians Wang Jianguo and Jiang Nan have noted, former industrial sites are the first places to be torn down during urban redevelopment in China.[38] Most Chinese see old factories as symbols of economic recession. In Jingdezhen, many former workers were nostalgic about their former work units, but they were also eager for the sites to be sold and redeveloped, so that they could realize at least a modest financial benefit from the sale.

Many outsiders believed that the influx of visiting ceramists to Jingdezhen was causing a 'creative renaissance'.[39] While international ceramic art was influencing the practice of students and faculty at the Jingdezhen Ceramics Institute, such claims attribute a far greater impact to the 'Jingdezhen drifters' than most locals would admit. Most of the city's ceramists continued to produce historic replicas, porcelain influenced by national painting, and art that resonated with Jingdezhen's imperial ceramics tradition. The ceramists who made art porcelain outside the academy mostly made a living by selling to brokers. The visiting artists did not employ large numbers of porcelain workers, and the work they provided was sporadic. The celebrity and economic rewards that came from such collaborations did not accrue locally. To take a well-known example, Ai Weiwei's massive sunflower seeds project employed hundreds more ceramists, for a much longer period, than international art projects typically did. The resulting porcelain sunflower seeds are arguably Ai's most famous work. By May 2012, sales of less than one-hundredth of the total sunflower seeds had already yielded more than $1.3 million dollars.[40] None of this money went to the Jingdezhen slip-casters or painters. Aficionados of the work cannot name a single ceramist who made the sunflower seeds, and some do not know that they were made in Jingdezhen, despite Ai's best efforts to publicize this fact.

From the time that the Song-dynasty emperor Zhenzong first took an interest in *qingbai* through to the late twentieth century, China's central government commissioned ceramics, managed porcelain production, and expended resources on innovation and technological improvements so long as the country had a functional regime. From 1949 to 1994, the role that the Chinese Communist Party played in Jingdezhen was consistent with past practice, albeit more intensive. When top leaders decided to privatize porcelain manufacturing, they made a dramatic break with Jingdezhen's history. From the late 1990s onward, commercial firms and private entrepreneurs, supply and demand, and market competition determined the success or failure of Jingdezhen's porcelain industry. Production was smaller-scale, comparatively homogenous in output, less stable, and employed fewer people. The population was more socially and economically stratified, with a few ceramists who drove imported cars and took international vacations, while other residents washed the family's laundry in the Chang River. In the absence of official regulation, copying and counterfeiting was rampant. While China's economy was growing and its middle class expanding, ordinary ceramists could 'make enough to eat' (*fen fan chi*). Even so, the less skilled and less well-connected participated in a 'race to the bottom' as they tried to make a living by cutting costs, self-exploiting their own labour, and producing ever more cheaply.

In 2013, the Ancient Kiln Folk Customs Museum put in a bid for UNESCO World Heritage Site status.[41] The application showed local officials', and to some extent the central government's, commitment to tourism and use of the imperial past for economic development. Jingdezhen's industrial past was vulnerable, as were many of those who had worked in its modern industrial enterprises. As officials in the former porcelain capital focused on selling culture, heritage, and experience, ceramists were left to figure out for themselves how to get work and get by. Without a major shift in government policy, the large-scale manufacturing that had made the city an international force, and supported hundreds of thousands of porcelain workers, would not return.

# Glossary

**Bad class labels** Landlord, capitalist, rich peasant. The Chinese Communist Party assigned class labels to all Chinese citizens based on their family's status prior to the Communist victory in 1949.

**Biscuit** Also bisque. Ceramic wares that have been fired but not glazed. The temperature for firing varies from 800 to 1,300 degrees Celsius depending on the clay body.

**Celadon** A general term used for green-glazed wares. The glaze may be transparent or opaque, and colours range from yellow-green to green-blue glaze. Celadons were made in both north and south China, from clays that were grey, brown, or white. The Song dynasty is well known for its celadons. The term comes from a character in the pastoral romance *L'Astrée* who wore ribbons of a grey-green tone.

**China stone** A rock that combines quartz and fine mica and is the main ingredient of Chinese porcelain. Also called petunse, porcelain stone, and 'little white bricks' (*bai dunzi*).

**Clay** A fine-grained earthy material that is plastic when wet and hardened when heated.

**Closed wares** Forms that cannot be made by throwing. For example, to make a square or hexagonal ware, clay must be rolled into slabs, which are then cut to the required sizes and luted together with slip. Closed wares are called *zhouqi* in Chinese.

**Coal gas** A by-product of coke production that can be used as fuel in tunnel kilns.

**Coke** A fuel made from coal that is used in blast furnaces to smelt iron.

**Cultural Revolution** A movement begun by Mao Zedong in 1966 to purge the Communist Party of capitalists and revisionists, train young people to be revolutionaries, and attack traditional culture. Also called the Great Proletarian Cultural Revolution and 'Ten Years of Chaos' (*shinian dongluan*). The Cultural Revolution caused the deaths of more than two million people, including scholars, artists, teachers, and scientists as well as political leaders.

**Deng Xiaoping** China's paramount leader from the late 1970s through the mid-1990s, and the architect of Reform and Opening, China's economic transition to capitalism. Deng was an early member of the Chinese Communist Party and held many high-level positions until he was accused of being a capitalist by Mao Zedong during the Cultural Revolution and sent to work in a tractor factory in Jiangxi Province. He returned to political office in 1973, and rose to become China's top leader in 1977.

**Earthenware** A porous ceramic, often reddish in colour, fired between 800 and 1,100 degrees Celsius. In Chinese earthenware are called *tao*.

**Feldspar** A group of minerals with alumina and silica in their chemistry; the single most abundant mineral on earth.

**Fire** To use heat to remove water from clay (or porcelain stone) and transform it from a pliable medium to a hard substance. Porcelain in Jingdezhen is fired at 1,300 degrees Celsius. Ceramics are usually fired in kilns, but can be fired in open pits.

**Glaze** A glassy coat of material that is often used to cover or partly cover a ceramic form. Glazes can be applied to low- or high-fired ceramics. In Jingdezhen, glazes are made from a special type of china stone (*youguo*) plus ash (usually limestone or calcium carbonate). Glazes can be transparent, opaque, and translucent. Transparent coloured glazes have colouring materials in solution. Opaque coloured glazes have crystals, materials in suspension. Translucent coloured glazes have materials in between solution and suspension. A potter can change glaze colours by varying kiln atmospheres; for example, a glaze that contains iron fires to a creamy ivory in oxidation, and a cold blue in reduction. Length of firing time also affects how a glaze looks.

**Great Leap Forward** A campaign begun by Mao Zedong in 1958 to modernize and industrialize China in a short space of time through grassroots mobilization. The bad economic policies implemented during the three years of the Great Leap contributed to a massive famine that killed tens of millions of citizens.

**Greenware** Unfired clay forms.

**Jigger-jolly machine** A machine for mechanized porcelain production. A jigger-jolly machine uses mechanized arms to enclose a spinning ball of clay in moulds and produce forms.

**Kaolin** White clays that are naturally low in iron oxides and fluxes.

**Kiln** The 'oven' in which ceramics are fired. A kiln usually has a firebox, chamber, and chimney connected by flues. In some kilns, pots are separated from direct contact with fire by placing them in saggars. Jingdezhen used wood-firing kilns until the mid-twentieth century, when officials switched the city to coal. Today, most Jingdezhen kilns fire with propane or coal gas.

**Levigate** A water-based process to remove unwanted materials from clays and china stone. In Jingdezhen, pulverized china stone is mixed with water in a pit and the mixture is allowed to settle. The unwanted materials sink and the fine china stone remains suspended and can be skimmed off and left to dry.

**Mao Zedong** China's premier leader from 1949 to 1976. Also called Chairman Mao. Mao served as chairman of the People's Republic of China from 1949 to 1959, and led the Chinese Communist Party from 1935 until his death. Mao was a Marxist theorist and army general as well as a political leader. China's two biggest political campaigns, the Great Leap Forward and the Cultural Revolution, are generally attributed to Mao.

**Mica** A silicate mineral that forms in layers and is a key component of china stone. Mica has a pearly lustre and is heat resistant.

**Nationalist Party** A political party that began as a revolutionary group to overthrow the Chinese monarchy. The Nationalist Party governed parts of China from 1928 to 1949 and subsequently ruled Taiwan under Chiang Kai-shek. Also called the Kuomintang or KMT.

**Overglaze enamels** Materials made from quartz, flux, and metal oxide pigment to add colour to the surface of pots. The materials are ground together, made into a suspension, and applied to ceramic bodies, which are then fired at low temperatures in a muffle kiln. Enamels are typically applied over a colourless or translucent glaze.

**Petunse** A term for china stone that may derive from the Chinese for 'little white bricks' (*bai dunzi*).

**Planned economy** An economy directed by the state, in which the government sets production goals, controls prices, and allocates resources.

**Porcelain** A hard, white-firing, translucent ceramic fired between 1,200 and 1,400 degrees Celsius. In Jingdezhen, porcelain is made from a naturally kaolinized china stone. Other parts of the world lack this material, and so porcelain in other places is created by adding whiteners to clay (for example, flint or animal bones).

**Propane** A colourless gas that can be used as a fuel.

**Qingbai** A thin-bodied porcelain celadon ware made in Jingdezhen during the Song dynasty, renowned for its jade-like quality.

**Raw glazing** Glazing greenware to make glassy ceramics through a single firing. For most of Jingdezhen's history ceramists have used raw glazing, rather than doing a bisque firing and a second glaze firing.

**Reduction atmosphere** An atmosphere in a kiln with low or reducing levels of oxygen. The atmosphere in a kiln affects the colour of the finished wares.

**Refractory** Able to withstand high heat.

**Round wares** Wares that can be thrown on a potter's wheel, including bowls, plates, dishes, etc.

**Saggar** A protective container in which greenware is fired, made of highly refractory material. Sometimes called a clay box or firing box. In Jingdezhen these were round, and made so they could be stacked atop one another in long columns. When ceramists or written sources talk about kilns that 'topple', they are referring to the columns of saggars toppling.

**Sherd** A fragment of a ceramic vessel.

**Slip** A watery liquid clay that can be used for surface decoration, slip-casting, or luting (connecting) clay forms together.

**Slip-casting** A way of producing clay forms, which involves pouring a watery liquid clay (slip) into a plaster of Paris mould. The mould absorbs liquid and a thin layer of clay (the cast) appears. The ceramist pours off excess slip and allows the clay to dry further, forming a clay body. Jingdezhen potters regarded slip-casting as mechanical production, although the process requires a lot of labour.

**Slurry** A semi-liquid mixture of particles suspended in water.

**Stoneware** A hard ceramic, often grey, yellow, or pale brown, fired between 1,200 and 1,300 degrees Celsius. Note that in Chinese no separate word exists for stoneware; all high-firing ceramics are *ci*.

**Taiping kiln** A ritual kiln fired during the Mid-Autumn Festival that marks the Taiping Rebellion in Jingdezhen.

**Throw** A method of making forms using a potter's wheel. In China, potters throw 'off the hump' or from the top of a large mound of clay. The potter compresses clay between her fingers atop a spinning wheel.

**Treaty ports** A term used to describe port cities in China that were opened to foreign trade by unequal treaties with foreign powers. The British established the first treaty ports in China at the end of the First Opium War. In many treaty ports, foreign countries established direct control over particular areas, known as concessions.

**Underglaze blue** Decoration with cobalt underneath a transparent glaze. Also called blue and white.

**Work unit** Under the socialist economy, a person's place of employment, which also provided him or her with housing, medical care, pensions, and other benefits. Work units offered recreational opportunities to their workers, such as choirs, dances, and movies. They were also a unit of social control: a worker needed permission from his or her work unit to marry or travel. For most citizens, a work unit was a lifetime employer and provider. Work units were differentiated by type: state enterprises were considered to be the best work units, with the most benefits, and small collective work units had the fewest benefits.

# Notes

## 1 The World's Most Famous Ceramics and the People Who Made Them

1 Euan McKirdy, '"Holy Grail" of Chinese Porcelain Nets Record Bid for Sotheby's', CNN.com, 10 April 2014, http://edition.cnn.com/2014/04/09/world/asia/sothebys-record-sale-chinese-porcelain/.
2 Rose Kerr and Nigel Wood, *Ceramic Technology, Science and Civilization in China* 5(12), ed. Joseph Needham (Cambridge, England: Cambridge University Press, 2004), 556.
3 John Carswell, *Blue and White: Chinese Porcelain Around the World* (Chicago, IL: Art Media Resources, 2000), 11–18. Not everyone agrees that the cobalt came from Persia. See Stacey Pierson, 'Precious Blue: Cobalt in Chinese Ceramic Glazes and Decoration,' *Taoci* 1 (2000), 36–7.
4 Kerr and Wood, *Ceramic Technology*, 692.
5 Other foreign visitors to Jingdezhen report the same tale. See, for example, Laura Imkamp, 'The Future of China: Jingdezhen Struggles to Maintain its Porcelain Fame,' *CNN Travel*, 20 July 2011, n.p.
6 Fang Lili, 景德鎮民窑 [*Jingdezhen's Folk Potteries*] (Beijing: People's Art Publisher, 2002), and 传统与变迁：景德镇旧民窑的田野调查 [*Traditions and Changes: Investigations into Jingdezhen's Old Folk Potteries*] (Nanchang: Jiangxi People's Publisher, 2000).
7 My own publications are a good example. See Maris Boyd Gillette, 'Labor and Precariousness in China's Porcelain Capital', *Anthropology of Work Review* 35 (2014a), 25–39; 'Documenting, Dramatizing, and Representing China's Porcelain World in Broken Pots Broken Dreams,' *Visual Anthropology Review* 30 (2014b), 38–49; 'Globalization and Deindustrialization in China's (Former) Porcelain Capital', *East Asia in the World: An Introductory Guide*, ed. Anne Prescott (Armonk, NY: M.E. Sharpe, Inc., 2015); 'Contemporary Jingdezhen's Porcelain Entrepreneurs', *Landscapes in Blue – Made in China: The Vase Project* (Lafayette, PA: Lafayette University, 2012), 44–53; 'Copying, Counterfeiting, and Capitalism in Jingdezhen's Porcelain Industry', *Modern China* 36 (2010), 367–403; *Broken Pots Broken Dreams* (film, Haverford, PA, 2009).
8 Nigel Wood, 'Plate Tectonics and Chinese Ceramics: New Insights into the Origins and Distribution of China's Ceramic Raw Material', *Taoci* 1 (2000), 15–24.

9 Jingdezhen City Government, 'A Brief Introduction of [sic] Jingdezhen, China'. Jingdezhen, http://eng.jdz.gov.cn/Brief/introduction/201112/t20111201_122542.htm, 2011a.

10 Wang Zongda and Yin Chengguo, eds., 現代景德鎮陶瓷經濟史 1949–1993 [*Modern Economic History of Jingdezhen 1949–1993*] (Beijing: Zhongguo Shuji Publications, 1994), 114–15.

11 A short film by Jackson Li showing this process is available at http://chinaclayart.com/.

12 Kerr and Wood, *Ceramic Technology*, 234. For a full discussion of kaolin in the Jingdezhen ceramic medium, see pages 229–37.

13 Jingdezhen City Records Compilation Committee, eds., 景德鎮市志略 [*Brief Records of Jingdezhen City*] (Shanghai: Chinese Dictionary Publishers, 1989), 311.

14 Robert Tichane, *Ching-te-chen: Views of a Porcelain City* (Painted Post, NY: The New York State Institute for Glaze Research, 1983). Tichane translates and compiles several accounts of Jingdezhen, ranging from a fourteenth-century description to his own travel account in 1981.

15 Kerr and Wood, *Ceramic Technology*, 206.

## 2 Creating a Porcelain Capital, Prehistory to 1785

1 Wu Xiaohong et al., 'Early Pottery at 20,000 Years Ago in Xianrendong Cave, China', *Science* 336 (2012), 1696–1700, doi: 10.1126/science.1218643.

2 Archaeologists have found early high-fired stonewares in multiple other sites, too. See Kerr and Wood, *Ceramic Technology*, 11–12.

3 Kerr and Wood, *Ceramic Technology*, 11, 18–19.

4 Chen Yuqian, ed., 景德鎮陶瓷文化概論 [*A Discussion of Jingdezhen's Ceramic Culture*] (Nanchang, Jiangxi: Jiangxi Higher Education Publishers, 2004), 24; R.L. Hobson, *Chinese Pottery and Porcelain* (New York: Dover Publications Inc., 1915 [1974]), 153.

5 Kerr and Wood, *Ceramic Technology*, 524.

6 Wang Qingzheng, *A Dictionary of Chinese Ceramics* (Singapore: Oriental Art Publications, 2002), 193.

7 Kerr and Wood, *Ceramic Technology*, 714–16.

8 See Kerr and Wood, *Ceramic Technology*, 347–57 for a basic overview of dragon kilns in south China.

9 Kerr and Wood, *Ceramic Technology*, 212–13.

10 Anne Gerritsen, 'Fragments of a Global Past: Ceramics Manufacture in Song-Yuan-Ming Jingdezhen', *Journal of the Economic and Social History of the Orient* 52 (2009), 133–4.

11 See John W. Chaffee, *Branches of Heaven: A History of the Imperial Clan of Sung China* (Cambridge, MA: Harvard University Press, 1999), 36–63; and Nap-yin Lau and Huang K'uan-chung, 'A New Type of Emperor: The Diffident Zhenzong, 997–1022', *The Cambridge History of China* 5(1), eds. Denis Twitchett and Paul Smith (Cambridge: Cambridge University Press, 2009), 260–78.

12 Kerr and Wood, *Ceramic Technology*, 185–6, 190; Michael Dillon, *A History of the Porcelain Industry in Jingdezhen* (Ph.D thesis, University of Leeds, 1976), 23.

13 Jiangxi Province Gazetteer Editorial Commission, 江西省陶瓷工業志 [*A Record of the Ceramics Industry in Jiangxi Province*] (Jiangxi: Jiangxi Province Gazetteer Publishing, 2005), 24.

14 See Kerr and Wood, *Ceramic Technology*, 19, for a translation and the original Chinese.

15 Ibid., 186.

16 Ibid., 201.

17 Chen, 景德鎮陶瓷文化概論, 211.

18 For a translation, see Tichane, *Ching-te-chen*, 43–8.

19 Dillon, *A History of the Porcelain Industry in Jingdezhen*, 24; Kerr and Wood, *Ceramic Technology*, 186–7, 209.

20 Jessica Harrison-Hall, 'The Ming Porcelain Industry at Jingdezhen 1368–1644', *Catalogue of Late Yuan and Ming Ceramics in the British Museum* (London: The British Museum Press, 2001), 21.

21 Robert Finlay, 'The Pilgrim Art: The Culture of Porcelain in World History', *Journal of World History* 9(2) (1998), 154–5.

22 Finlay, 'The Pilgrim Art', 156.

23 Kerr and Wood, *Ceramic Technology*, 186.

24 See Gerritsen, 'Fragments of a Global Past: Ceramics Manufacture in Song-Yuan-Ming Jingdezhen'.

25 Gerritsen, 'Fragments of a Global Past: Ceramics Manufacture in Song-Yuan-Ming Jingdezhen', 119.

26 Many ceramists used this phrase to describe porcelain production during my field research. See also Chen, 景德鎮陶瓷文化概論, 31.

27 See Dillon, *A History of the Porcelain Industry in Jingdezhen*, 26, 30, 44; Kerr and Wood, *Ceramic Technology*, 191–2, 641.

28 Kerr and Wood, *Ceramic Technology*, 209.

29 Harrison-Hall, 'The Ming Porcelain Industry at Jingdezhen 1368–1644', 24.

30 Harrison-Hall, 'The Ming Porcelain Industry at Jingdezhen 1368–1644', 24; Michael Dillon, 'Jingdezhen as a Ming Industrial Center', *Key Papers on Chinese Economic History up to 1949* (Folkestone: Global Oriental, 2007), 286; Kerr and Wood, *Ceramic Technology*, 209.

31 Archibald Dooley Brankston, *Early Ming Wares at Jingdezhen* (Peking: H. Vetch, 1938), 62–3.
32 Kerr and Wood, *Ceramic Technology*, 212.
33 Kerr and Wood, *Ceramic Technology*, 235–7.
34 Kerr and Wood, *Ceramic Technology*, 279–81, 469.
35 Kerr and Wood, *Ceramic Technology*, 365–78.
36 Kerr and Wood, *Ceramic Technology*, 371–2.
37 Kerr and Wood, *Ceramic Technology*, 619, 707.
38 Kerr and Wood, *Ceramic Technology*, 676.
39 Kerr and Wood, *Ceramic Technology*, 564.
40 Kerr and Wood, *Ceramic Technology*, 573.
41 Kerr and Wood, *Ceramic Technology*, 202.
42 Luo Xuezheng, '景德镇明清官窑落选瓷的变卖与仿制' ['The Sale and Reproduction of Imperial Kiln Seconds During the Ming and Qing Dynasties'], 文物天地 [*Wenwu Tiandi*] 12 (2004): 29.
43 Kerr and Wood, *Ceramic Technology*, 197–200, 207.
44 I met several locals who described how they dug for sherds at the former imperial manufactory before the government excavated the site. See also Luo Xuezheng, '景德镇明清官窑落选瓷的变卖与仿制', 28.
45 Luo Xuezheng, '景德镇明清官窑落选瓷的变卖与仿制', 28.
46 See Kerr and Wood, *Ceramic Technology*, 198; Jiangxi Gazetteer Editorial Commission, 江西省陶瓷工業志, 24; and Chen, 景德鎮陶瓷文化概論, 22.
47 Harrison-Hall, 'The Ming Porcelain Industry at Jingdezhen 1368–1644', 22.
48 Kerr and Wood, *Ceramic Technology*, 188. For the following figures concerning sixteenth-century production see page 199, table 33. The stone inscription is described on page 188.
49 Dillon, 'Jingdezhen as a Ming Industrial Center', 286; Wanda Garnsey and Rewi Alley, *China: Ancient Kilns and Modern Ceramics: A Guide to the Potteries* (Canberra: ANU Press, 1983), 107; Harrison-Hall, 'The Ming Porcelain Industry at Jingdezhen 1368–1644', 24.
50 See Fang Lili, 景德鎮民窰, 145, 147, 171–2, 176.
51 Kerr and Wood, *Ceramic Technology*, 445.
52 Dillon, 'Jingdezhen as a Ming Industrial Center', 286–7.
53 Jiangxi Light Industry Bureau Ceramics Research Center, 景德鎮陶瓷史稿 [*A Draft History of Jingdezhen Ceramics*] (Beijing: Sanlian Publishers, 1959), 239. See also Kerr and Wood, *Ceramic Technology*, 210. The original source is the *Mingshilu* (*Ming Historical Records*).
54 Fang, 景德鎮民窰, 145; Kerr and Wood, *Ceramic Technology*, 212.
55 Fang Lili, 景德鎮民窰, 147, 185–6, 199–200, 203, 220, 240, etc.; Jiang, 景德鎮瓷業史 [*A History of Jingdezhen's Porcelain Industry*] (Shanghai: Zhonghua Shuju, 1936), 183–5.

56 See examples in Fang, 景德鎮民窰, 185–6, 232, 236–7, etc.; and Jiang Siqing, 景德鎮瓷業史, 179.

57 See Wang Zongda, Yin Chengguo, eds., 現代景德鎮陶瓷經濟史 1949–1993, 73, 109.

58 Quoted from Chen, 景德鎮陶瓷文化概論, 301. Ying Zilin, ed., 日用瓷工藝 [*The Art of Daily Use Porcelain*] (Nanchang, Jiangxi: Nanchang Linyun Publishers, 1988), 4, writes that the term 'porcelain capital' dates from the Ming dynasty.

59 See Finlay, 'The Pilgrim Art', 142, 168–73.

60 Finlay, 'The Pilgrim Art', 174–5; Kerr and Wood, *Ceramic Technology*, 747–67.

61 Craig Clunas, 'The Cost of Ceramics and the Cost of Collecting Ceramics in the Ming Period', *Bulletin of the Oriental Ceramics Society of Hong Kong* 8, 1986–1988 (1991), 50; Hobson, *Chinese Pottery and Porcelain*, 95.

62 See Liang Miaotai, 明清景德镇城市经济研究 [*Research on Jingdezhen's Urban Economy During the Ming and Qing Dynasties*] (Nanchang: Jiangxi People's Publisher, 1991), 221.

63 Jingdezhen City Records Compilation Committee, 景德镇市志略, 27–33.

64 Dillon, *A History of the Porcelain Industry in Jingdezhen*, 42.

65 Ibid., 42; Kerr and Wood, *Ceramic Technology*, 210–11.

66 Kerr and Wood, *Ceramic Technology*, 188, 210–11; Rose Kerr, *Chinese Ceramics: Porcelain of the Qing Dynasty 1644–1911* (London: Victoria and Albert Museum, 1986), 18; Dillon, *A History of the Porcelain Industry in Jingdezhen*, 44–5.

67 Kerr and Wood, *Ceramic Technology*, 200.

68 Kerr and Wood, *Ceramic Technology*, 211.

69 As reported by d'Entrecolles, whose letters to the French government are translated in Tichane, *Ching-te-chen*, 55–7. See also Chen Yuqian, 景德鎮陶瓷文化概論, 22; and Dillon, *A History of the Porcelain Industry in Jingdezhen*, 47–8. I quote from d'Entrecolles 1712 letter below.

70 Dillon, *A History of the Porcelain Industry in Jingdezhen,* 50–1.

71 Stanley Wright, *Kiangsi Native Trade and Its Taxation* (Shanghai: n.p., 1920), 21.

72 Luo Xuezheng, '景德镇仿古瓷的历史与现状' ['The History and Current Situation of Antique Reproductions in Jingdezhen'], 美術 [*Meishu*] 7(2005): 100; Kerr and Wood, *Ceramic Technology*, 207, 602, 608.

73 Luo Xuezheng, '景德镇仿古瓷的历史与现状', 101.

74 Luo Xuezheng, '景德镇明清官窑落选瓷的变卖与仿制', 29.

75 Chen, 景德鎮陶瓷文化概論, 31–6.

76 Kerr and Wood, *Ceramic Technology*, 686.

77 Fang Lili's 景德镇民窑 describes Jingdezhen's specialized labour in minute detail. While her sources are mostly from the late Qing and Republican period, she writes that her research speaks to Jingdezhen's organization from the Ming onwards, as the porcelain industry was conservative and handicraft production stable from the sixteenth through the early twentieth centuries. See pages 6–7.

78 Kerr and Wood, *Ceramic Technology*, 200.
79 Gerritsen, 'Fragments of a Global Past: Ceramics Manufacture in Song-Yuan-Ming Jingdezhen', 142.
80 According to the 1815 *Jingdezhen Taolu*, cited in Fang, 景德镇民窑, 160. Fang describes a typical round-ware workshop on pages 160–1.
81 Kerr and Wood, *Ceramic Technology*, 448–50.
82 Fang, 景德镇民窑, 153.
83 Kerr and Wood, *Ceramic Technology*, 213; Fang, 景德镇民窑, 152.
84 Fang, 景德镇民窑, 171–2.
85 Kerr and Wood, *Ceramic Technology*, 374; Fang, 景德镇民窑, 173.
86 Fang, 景德镇民窑, 168.
87 Chen, 景德镇陶瓷文化概论, 25.
88 Fang, 景德镇民窑, 173.
89 Fang, 景德镇民窑, 175. Foshan is known for producing temple tiles and figurines.
90 Kerr and Wood, *Ceramic Technology*, 646.
91 Fang, 景德镇民窑, 175–6.
92 Chen, 景德镇陶瓷文化概论, 34–6; Fang, 景德镇民窑, 175–7.
93 For information about brush-makers, see Fang, 景德镇民窑, 184. For information about flux producers, see Fang, 景德镇民窑, 185.
94 Kerr and Wood, *Ceramic Technology*, 226–8.
95 Chen, 景德镇陶瓷文化概论, 33–4.
96 Fang, 景德镇民窑, 196.
97 Dillon, *A History of the Porcelain Industry in Jingdezhen*, 164, 180; Fang, 景德镇民窑, 218, 220.
98 Dillon, *A History of the Porcelain Industry in Jingdezhen*, 51.
99 Zhou Jianshu, 景德镇史话 [*Speaking of Jingdezhen's History*], (Shanghai: People's Publisher, 1989), 195. See also page 181.

## 3 Decline and Disarray, 1780–1948

1 See Jiangxi Light Industry Bureau Ceramics Research Center, 景德鎮陶瓷史稿, pages 275–9, for the original text of Du's report. This translation is mine. I reorganized portions of the report to put all Du's comments on specific aspects of the industry adjacent to one another.
2 Nanchang, the provincial capital of Jiangxi, is 270 km from Jingdezhen.
3 Jiang, 景德鎮瓷業史, 165.
4 Jiangxi Province Gazetteer Editorial Commission, eds, 江西省陶瓷工業志, 20.
5 Dillon, *A History of the Porcelain Industry in Jingdezhen*, 51; and Kerr and Wood, *Ceramic Technology*, 197.

6 Jiangxi Province Gazetteer Editorial Commission, 江西省陶瓷工業志, 25.

7 Dillon, *A History of the Porcelain Industry in Jingdezhen*, 54; Jiang, 景德鎮瓷業史, 157; Kerr and Wood, *Ceramic Technology*, 189.

8 Georges F. Scherzer, 'A Trip to Ching-te-chen', *Ching-Te-Chen: Views of a Porcelain City*, ed. Robert Tichane (Painted Post, NY: New York State Institute for Glaze Research, 1900 [1983]), 191.

9 Kerr and Wood, *Ceramic Technology*, 197–201.

10 Kamal Sheel, *Peasant Society and Marxist Intellectuals in China: Fang Zhimin and the Origin of a Revolutionary Movement in the Xinjiang Region* (Princeton, NJ: Princeton University Press, 1989), 62–3; Jiangxi Light Industry Bureau Ceramics Research Center, 景德鎮陶瓷史稿, 268; Zhou, 景德鎮史話, 141.

11 Jiangxi Light Industry Bureau Ceramics Research Center, 景德鎮陶瓷史稿, 269; Zhou, 景德鎮史話, 141.

12 Albert Feuerwerker, 'Economic Trends in the Late Ch'ing Empire, 1870–1911', *Cambridge History of China: Late Ch'ing*, volume 11, part 2, eds. John K. Fairbank and Kwang-Ching Liu (Cambridge: Cambridge University Press, 1980), 19; Dillon, *A History of the Porcelain Industry in Jingdezhen*, 56.

13 Walter Mallory, *China: The Land of Famine* (New York: American Geographical Society, 1926), 42–58; Marianne Bastid-Bruguiere, 'Currents of Social Change', *Cambridge History of China: Late Ch'ing*, volume 11, part 2, eds. John K. Fairbank and Kwang-Ching Liu (Cambridge: Cambridge University Press, 1980), 586, 596.

14 W.J. Clennell, 'A Report on a Journey in the Interior of Kiangsi', *Ching-Te-Chen: Views of a Porcelain City*, ed. Robert Tichane (Painted Post, NY: New York State Institute for Glaze Research, 1905 [1983]), 325–6; Jiang, 景德鎮瓷業史, 162.

15 Chen, 景德鎮陶瓷文化概論, 212; Jiang, 景德鎮瓷業史, 162; Zhou, 景德鎮史話, 139.

16 Mallory, *China: The Land of Famine*, 79; see also Jiang, 景德鎮瓷業史, 163–5.

17 Dillon, *A History of the Porcelain Industry in Jingdezhen*, 57–8.

18 Chen, 景德鎮陶瓷文化概論, 214; Jiang, 景德鎮瓷業史, 198; Zhou, 景德鎮史話, 153–60.

19 Dillon, *A History of the Porcelain Industry in Jingdezhen*, 61.

20 Jiang, 景德鎮瓷業史, 199.

21 Jiang, 景德鎮瓷業史, 160–1; Jiangxi Light Industry Bureau Ceramics Research Center, 景德鎮陶瓷史稿, 265.

22 Jiang, 景德鎮瓷業史, 166; Zhou, 景德鎮史話, 139.

23 Clennell, 'A Report on a Journey in the Interior of Kiangsi', 330–1; and Fang, 景德鎮民窰, 175.

24 Jiangxi Light Industry Bureau Ceramics Research Center, 景德鎮陶瓷史稿, 286; and Jiang, 景德鎮瓷業史, 182.

25 Frank B. Lenz, 'The World's Ancient Porcelain Center', *The National Geographic Magazine* 38(5) (1920), 405.
26 Fang, 景德鎮民窰, 240.
27 Jiang, 景德鎮瓷業史, 189.
28 Fang, 景德鎮民窰, 239.
29 Harry Franck, *Roving Through Southern China* (New York: The Century Company, 1925), 377; Jiang, 景德鎮瓷業史, 188; John Knight Shyrock, 'Kingtehchen The Porcelain City', *Asia* (November 1920), 997.
30 Jiang, 景德鎮瓷業史, 158; Fang, 景德鎮民窰, 240.
31 Fang, 景德鎮民窰, 214.
32 Fang, 景德鎮民窰, 239.
33 Clennell, 'A Report on a Journey in the Interior of Kiangsi', 328; Lenz, 'The World's Ancient Porcelain Center', 405; Dillon, *A History of the Porcelain Industry in Jingdezhen*, 178; Fang, 景德鎮民窰, 239–40.
34 Michael Dillon, 'Fang Zhimin, Jingdezhen, and the Northeast Jiangxi Soviet: Tradition, Revolution, and Civil War in a Pottery Town,' *Modern Asian Studies* 26(3) (1992), 575, Jiangxi Light Industry Bureau Ceramics Research Center, 景德鎮陶瓷史稿, 339–40.
35 Sources for Table 2 are: Jiangxi Light Industry Bureau Ceramics Research Center, 景德鎮陶瓷史稿, 239, 269, 339–40, 342–3, 345–6; Fang, 景德鎮民窰, 176, 220–1, 239, 241; Dillon, *A History of the Porcelain Industry in Jingdezhen*, 52, 161–2; Dillon, 'Fang Zhimin, Jingdezhen, and the Northeast Jiangxi Soviet: Tradition, Revolution, and Civil War in a Pottery Town', 575–82; Jiang, 景德鎮瓷業史, 158, 189, 164; Wright, *Kiangsi Native Trade and Its Taxation*, 193–4; Franck, *Roving Through Southern China*, 364; Zhou, 景德鎮史話, 160, 166–7, 171–3.
36 See Jiangxi Light Industry Bureau Ceramic Research Center, 景德鎮陶瓷史稿, 237, 238, 334; Jiang, 景德鎮瓷業史, 188; Fang, 景德鎮民窰, 240.
37 Dillon, 'Fang Zhimin, Jingdezhen, and the Northeast Jiangxi Soviet: Tradition, Revolution, and Civil War in a Pottery Town', 578–84; Jiangxi Light Industry Bureau Ceramic Research Center, 景德鎮陶瓷史稿, 341–4.
38 Franck, *Roving Through Southern China*, 108.
39 Jingdezhen City Records Compilation Committee, eds., 景德鎮市志略, 40.
40 Wang and Yin, 現代景德鎮陶瓷經濟史, 21, 25; Jiangxi Province Gazetteer Editorial Commission, 江西省陶瓷工業志, 25.

## 4 Production and Politics, 1949–1972

1 The troops that arrived were the Second Field Army, Fifth Corps, Seventeenth Armed Division. Detailed information about the PLA's arrival in Jingdezhen, and

the flight of the local government, can be found in Jingdezhen City Records Compilation Committee, 景德鎮市志略, 307. See also Jingdezhen City Records Compilation Committee, 景德鎮市志略, 40–2, 222; Ding Weizhi, ed., 景德镇卷 [*A Record of Jingdezhen*] (Beijing: China Encyclopedia Publisher, 1996), 285; Wang and Yin, 現代景德鎮陶瓷經濟史, 5, 22, 24–5, 27.

2 All biographical information in this and subsequent chapters comes from my ethnographic research in Jingdezhen, 2003–2010, unless otherwise noted. Readers can watch Feng Shangsong talk about his experiences in my film *Broken Pots Broken Dreams*. In addition to my ethnographic interviews with Feng, I also use his short unpublished autobiography entitled '我的七十年回顾' [My Memories at 70] (Jingdezhen: n.p., 2008).

3 Jiangxi Province Gazetteer Editorial Commission, 江西省陶瓷工業志, 20; Wang and Yin, 現代景德鎮陶瓷經濟史, 29–30, 59.

4 Jiangxi Light Industry Bureau Ceramic Research Center, 景德鎮陶瓷史稿, 350.

5 Ding, 景德镇卷, 60.

6 Ding, 景德镇卷, 142–3, 279; Zhou, 景德鎮史話, 176.

7 Jingdezhen City Records Compilation Committee, 景德鎮市志略, 307.

8 Wang and Yin, 現代景德鎮陶瓷經濟史, 34; Jingdezhen City Records Compilation Committee, 景德鎮市志略, 38.

9 Wang and Yin, 現代景德鎮陶瓷經濟史, 70–1, 78–9; Zhou, 景德鎮史話, 176; Ding, 景德镇卷, 170; Jingdezhen City Records Compilation Committee, 景德鎮市志略, 45, 51; Jiangxi Light Industry Bureau Ceramics Research Center, 景德鎮陶瓷史稿, 353.

10 Jiangxi Light Industry Bureau Ceramic Research Center, 景德鎮陶瓷史稿, 351; Wang and Yin, 現代景德鎮陶瓷經濟史, 25.

11 Jingdezhen City Records Compilation Committee, 景德鎮市志略, 307.

12 Wang and Yin, 現代景德鎮陶瓷經濟史, 50.

13 Wang and Yin, 現代景德鎮陶瓷經濟史, 57, 80, 221; Zhou, 景德鎮史話, 176; Ding, 景德鎮卷, 138.

14 Wang and Yin, 現代景德鎮陶瓷經濟史, 31, 48.

15 Ibid., 35, 60, 90–1.

16 Jingdezhen City Records Compilation Committee, 景德鎮市志略, 222; Ding, 景德镇卷, 318.

17 Wang and Yin, 現代景德鎮陶瓷經濟史, 71.

18 Jingdezhen City Records Compilation Committee, 景德鎮市志略, 308–309, Ding, 景德镇卷, 143.

19 Wang and Yin, 現代景德鎮陶瓷經濟史, 32.

20 Wang and Yin, 現代景德鎮陶瓷經濟史, 63–5, 86–9; Jiangxi Province Light Industry Bureau Ceramic Research Center, 景德鎮陶瓷史稿, 371; Jingdezhen City Records Compilation Committee, 景德鎮市志略, 309.

21 Wang and Yin, 現代景德鎮陶瓷經濟史, 73, 109.

22 Jingdezhen City Records Compilation Committee, 景德鎮市志略, 309; Wang and Yin, 現代景德鎮陶瓷經濟史, 151–2.

23 The official population in December 1949 was 192,700 residents. By 1952 the number of residents was probably larger. See Ding, 景德镇卷, 284.

24 See Maurice Meisner, *Mao's China and After: A History of the People's Republic* (New York: Free Press, 1999), 72.

25 Wang and Yin, 現代景德鎮陶瓷經濟史, 148–50.

26 Ibid., 71, 73, 95–6.

27 Ibid., 38.

28 Ibid., 71.

29 Jingdezhen City Records Compilation Committee, 景德鎮市志略, 309.

30 Jiangxi Province Light Industry Bureau Ceramic Research Center, 景德鎮陶瓷史稿, 355, 380.

31 Ibid., 386.

32 Ding, 景德鎮卷, 279; Wang and Yin, 現代景德鎮陶瓷經濟史, 207.

33 Zhou, 景德鎮史話, 176–7; Jiangxi Province Light Industry Bureau Ceramic Research Center, 景德鎮陶瓷史稿, 370.

34 Wang and Yin, 現代景德鎮陶瓷經濟史, 48, 53–4.

35 Jiangxi Province Light Industry Bureau Ceramic Research Center, 景德鎮陶瓷史稿, 375.

36 Jingdezhen City Records Compilation Committee, 景德鎮市志略, 45, 311; Jiangxi Province Light Industry Bureau Ceramic Research Center, 景德鎮陶瓷史稿, 377; Wang and Yin, 現代景德鎮陶瓷經濟史, 103; Ding, 景德鎮卷, 139–40; and Zhou, 景德鎮史話, 177.

37 Wang and Yin, 現代景德鎮陶瓷經濟史, 125–6, 140–2, 240; Zhou, 景德鎮史話, 179.

38 Wang and Yin, 現代景德鎮陶瓷經濟史, 110–12, 145; Jiangxi Province Light Industry Bureau Ceramic Research Center, 景德鎮陶瓷史稿, 281, 367, 383; Jingdezhen City Porcelain Industry Wage Reform Office, 景德鎮市瓷業工資改革資料彙集 [Compilation of Jingdezhen City Porcelain Industry Wage Reform Data] (n.p.: n.p, 1956), n.p.

39 Wang and Yin, 現代景德鎮陶瓷經濟史, 145; Jiangxi Province Light Industry Bureau Ceramic Research Center, 景德鎮陶瓷史稿, 382.

40 Fang, 景德镇民窑, 227.

41 Ibid., 158–9, 173.

42 Jiangxi Province Light Industry Bureau Ceramic Research Center, 景德鎮陶瓷史稿, 373.

43 Jingdezhen City Records Compilation Committee, 景德鎮市志略, 310; Ding, 景德鎮卷, 240.

44 Jingdezhen City Records Compilation Committee, 景德鎮市志略, 310.

45 Jingdezhen City Records Compilation Committee, 景德鎮市志略, 39; Zhou, 景德鎮史話, 188–9.

46 Jingdezhen City Records Compilation Committee, 景德鎮市志略, 222; Wang and Yin, 現代景德鎮陶瓷經濟史, 212; Zhou, 景德鎮史話, 188–9.

47 Jingdezhen City Records Compilation Committee, 景德鎮市志略, 40–1, 311.

48 Wang and Yin, 現代景德鎮陶瓷經濟史, 132.

49 Jingdezhen City Records Compilation Committee, 景德鎮市志略, 311.

50 Jiangxi Province Light Industry Bureau Ceramic Research Center, 景德鎮陶瓷史稿, 388–392, Jingdezhen City Records Compilation Committee, 景德鎮市志略, 311.

51 Zhou, 景德鎮史話, 176; Wang and Yin, 現代景德鎮陶瓷經濟史, 137; Jingdezhen City Records Compilation Committee, 景德鎮市志略, 311.

52 Wang and Yin, 現代景德鎮陶瓷經濟史, 115–16; Jiangxi Province Light Industry Bureau Ceramic Research Center, 景德鎮陶瓷史稿, 358–9.

53 Jingdezhen City Records Compilation Committee, 景德鎮市志略, 45, 310; Wang and Yin, 現代景德鎮陶瓷經濟史, 115; Ding, 景德鎮卷, 174.

54 Jingdezhen City Records Compilation Committee, 景德鎮市志略, 40, 49, 311; Ding, 景德鎮卷, 177; cf. Zhou, 景德鎮史話, 182.

55 Wang and Yin, 現代景德鎮陶瓷經濟史, 216.

56 Jiangxi Province Light Industry Bureau Ceramic Research Center, 景德鎮陶瓷史稿, 356; Zhou, 景德鎮史話, 178.

57 Jiangxi Province Light Industry Bureau Ceramic Research Center, 景德鎮陶瓷史稿, 357, 386–7, 397, 400; Jingdezhen City Records Compilation Committee, 景德鎮市志略, 310.

58 Jingdezhen City Records Compilation Committee, 景德鎮市志略, 31–2, 310–11. According to city records, the Chang River flooded 37 times between 795 and 1985.

59 Jiangxi Province Light Industry Bureau Ceramic Research Center, 景德鎮陶瓷史稿, 367.

60 Wang and Yin, 現代景德鎮陶瓷經濟史, 125–6, 135, 137, 140–2.

61 Jingdezhen City Records Compilation Committee, 景德鎮市志略, 221.

62 Ding, 景德鎮卷, 69.

63 Ding, 景德鎮卷, 140.

64 Wang and Yin, 現代景德鎮陶瓷經濟史, 186–7.

65 Ding, 景德鎮卷, 69; Wang and Yin, 現代景德鎮陶瓷經濟史, 138.

66 Frank Dikötter, *Mao's Great Famine: The History of China's Most Devastating Catastrophe, 1958–1962* (New York: Walker and Co., 2010), x. Dikötter's monograph is the most detailed study of the Great Leap Forward famine to date, informed by archives that have only been available for the past 15 years. For a succinct description of the Great Leap Forward in context, see Nicholas R. Lardy, 'The Chinese Economy Under Stress, 1958–1965', *The Cambridge History of China* 14, eds.

Roderick MacFarquhar and John King Fairbank (Cambridge: Cambridge University Press), 367–78.

67 Dikötter, *Mao's Great Famine*, 10–14.

68 Wang and Yin, 現代景德鎮陶瓷經濟史, 184.

69 Wang and Yin, 現代景德鎮陶瓷經濟史, 221; Jingdezhen City Records Compilation Committee, 景德鎮市志略, 312; Zhou, 景德鎮史話, 178.

70 Wang and Yin, 現代景德鎮陶瓷經濟史, 236.

71 Jingdezhen City Records Compilation Committee, 景德鎮市志略, 313.

72 Wang and Yin, 現代景德鎮陶瓷經濟史, 195–7, 212; Ding, 景德鎮卷, 174–5, 287; Jingdezhen City Records Compilation Committee, 景德鎮市志略, 55.

73 Jingdezhen City Records Compilation Committee, 景德鎮市志略, 312–13.

74 Dikötter, *Mao's Great Famine*, 170.

75 Wang and Yin, 現代景德鎮陶瓷經濟史, 223.

76 Ibid., 205.

77 Jingdezhen City Records Compilation Committee, 景德鎮市志略, 313.

78 Ibid.

79 Jingdezhen City Records Compilation Committee, 景德鎮市志略, 313; Wang and Yin, 現代景德鎮陶瓷經濟史, 188.

80 Wang and Yin, 現代景德鎮陶瓷經濟史, 223–44.

81 Wang and Yin, 現代景德鎮陶瓷經濟史, 206; Ding, 景德鎮卷, 144.

82 Wang and Yin, 現代景德鎮陶瓷經濟史, 231.

83 Ibid., 195, 206, 236–7.

84 Ibid., 238.

85 Jingdezhen City Records Compilation Committee, 景德鎮市志略, 314.

86 Wang and Yin, 現代景德鎮陶瓷經濟史, 233–5.

87 Ibid., 189.

88 Ding, 景德鎮卷, 287; Wang and Yin, 現代景德鎮陶瓷經濟史, 245.

89 Wang and Yin, 現代景德鎮陶瓷經濟史, 224–5, 263.

90 Lardy, 'The Chinese Economy Under Stress, 1958–1965', 367–72; Dikötter, *Mao's Great Famine*, 292–305.

91 Jingdezhen City Records Compilation Committee, 景德鎮市志略, 314.

92 Lardy, 'The Chinese Economy Under Stress, 1958–1965', 386–8; Wang and Yin, 現代景德鎮陶瓷經濟史, 246–7, 251.

93 Wang and Yin, 現代景德鎮陶瓷經濟史, 247; Ding, 景德鎮卷, 287.

94 Ding, 景德鎮卷, 527; Wang and Yin, 現代景德鎮陶瓷經濟史, 246.

95 Wang and Yin, 現代景德鎮陶瓷經濟史, 223, 242.

96 Wang and Yin, 現代景德鎮陶瓷經濟史, 256–7; Ding, 景德鎮卷, 171.

97 Jingdezhen City Records Compilation Committee, 景德鎮市志略, 315–16; Zhou, 景德鎮史話, 179, 182.

98 Lardy, 'The Chinese Economy Under Stress, 1958–1965', 386–91.

99 Martin Whyte, 'Urban Life in the People's Republic', *The Cambridge History of China* 15, eds. Roderick MacFarquhar and John King Fairbank (Cambridge: Cambridge University Press, 1991), 710.
100 Ding, 景德鎮卷, 144, 146; Jingdezhen City Records Compilation Committee, 景德鎮市志略, 316.
101 Ding, 景德鎮卷, 145, 287.
102 Jingdezhen City Records Compilation Committee, 景德鎮市志略, 316–17; Ding, 景德鎮卷, 287.
103 Jingdezhen City Records Compilation Committee, 景德鎮市志略, 59, 317; Wang and Yin, 現代景德鎮陶瓷經濟史, 248–50, 278, 291.
104 Wang and Yin, 現代景德鎮陶瓷經濟史, 254–55, 314.
105 Ibid., 263–7.
106 Ibid., 272, 276, 311.
107 Jingdezhen City Records Compilation Committee, 景德鎮市志略, 316; Zhou, 景德鎮史話, 190.
108 Jingdezhen City Records Compilation Committee, 景德鎮市志略, 316.
109 Ibid., 318.
110 For an account of the start of the Cultural Revolution, see Stephen Uhalley, 'The Cultural Revolution and the Attack on the "Three Family Village"', *The China Quarterly* 27 (1966): 149–61.
111 Jingdezhen City Records Compilation Committee, 景德鎮市志略, 318.
112 Wang and Yin, 現代景德鎮陶瓷經濟史, 297–8, 308, 319.
113 Wang and Yin, 現代景德鎮陶瓷經濟史, 292.
114 Jingdezhen City Records Compilation Committee, 景德鎮市志略, 318; Wang and Yin, 現代景德鎮陶瓷經濟史, 292.
115 Jingdezhen City Records Compilation Committee, 景德鎮市志略, 319.
116 Jingdezhen City Records Compilation Committee, 景德鎮市志略, 319; Wang and Yin, 現代景德鎮陶瓷經濟史, 295.
117 Wang and Yin, 現代景德鎮陶瓷經濟史, 298; Jingdezhen City Records Compilation Committee, 景德鎮市志略, 319; Ding, 景德鎮卷, 144, 146, 238.
118 Jingdezhen City Records Compilation Committee, 景德鎮市志略, 319–20; Wang and Yin, 現代景德鎮陶瓷經濟史, 292.
119 Wang and Yin, 現代景德鎮陶瓷經濟史, 293; Jingdezhen City Records Compilation Committee, 景德鎮市志略, 320.
120 Jingdezhen City Records Compilation Committee, 景德鎮市志略, 320; Wang and Yin, 現代景德鎮陶瓷經濟史, 303–4.
121 Jingdezhen City Records Compilation Committee, 景德鎮市志略, 323.
122 Jingdezhen City Records Compilation Committee, 景德鎮市志略, 320; Ding, 景德鎮卷, 289.
123 Ding, 景德鎮卷, 289.

124 Jingdezhen City Records Compilation Committee, 景德鎮市志略, 321; Wang and Yin, 現代景德鎮陶瓷經濟史, 304.

125 Jingdezhen City Records Compilation Committee, 景德鎮市志略, 323.

126 Wang and Yin, 現代景德鎮陶瓷經濟史, 299–301, 311; Jingdezhen City Records Compilation Committee, 景德鎮市志略, 321.

127 Wang and Yin, 現代景德鎮陶瓷經濟史, 303–4; Jingdezhen City Records Compilation Committee, 景德鎮市志略, 321.

128 Jingdezhen City Records Compilation Committee, 景德鎮市志略, 321.

129 Jingdezhen City Records Compilation Committee, 景德鎮市志略, 322.

130 Wang and Yin, 現代景德鎮陶瓷經濟史, 316–17. Details about the end of the Cultural Revolution in Jingdezhen can be found in Wang and Yin, 現代景德鎮陶瓷經濟史, 316–17, 319, 330, 456; and Jingdezhen City Records Compilation Committee, 景德鎮市志略, 322–4.

## 5 Dual-Track Porcelain, 1973–1993

1 Frank Cosentino, *The Boehm Journey to Ching-te-chen, China, Birthplace of Porcelain* (Trenton, NJ: Edward Marshall Boehm, Inc., 1976), 1, 6–9, 92. See also Wang and Yin, 現代景德鎮陶瓷經濟史, 331. For information on the expulsion of foreign missionaries from Jingdezhen, see Jingdezhen City Records Compilation Committee, 景德鎮市志略, 59.

2 Cosentino, *The Boehm Journey to Ching-te-chen*, 121.

3 I take the following description of the trip from Cosentino, *The Boehm Journey to Ching-te-chen*, 91–126, unless otherwise noted.

4 Wang and Yin, 現代景德鎮陶瓷經濟史, 300–1; Jingdezhen City Records Compilation Committee, 景德鎮市志略, 323.

5 Cosentino calls the institute the Ching-te-Chen Research and Design Institute and does not provide the Chinese name. Jingdezhen has had several ceramics research institutes, some of which were set up in the early 1950s (see Wang and Yin, 現代景德鎮陶瓷經濟史, 117). Some of the research centres' names have changed over time. Today some of these research centres have been absorbed by the Jingdezhen Ceramics Institute (which reopened in 1973).

6 Cosentino, *The Boehm Journey to Ching-te-chen*, 123.

7 Ibid., 156.

8 Wang and Yin, 現代景德鎮陶瓷經濟史, 362.

9 Ding, 景德镇卷, 289; Jingdezhen City Records Compilation Committee, 景德鎮市志略, 324; Wang and Yin, 現代景德鎮陶瓷經濟史, 317–18.

10 Jingdezhen City Records Compilation Committee, 景德鎮市志略, 328.

11 Wang and Yin, 現代景德鎮陶瓷經濟史, 317–18, 342; Jingdezhen City Records Compilation Committee, 景德鎮市志略, 325.

12 Wang and Yin, 現代景德鎮陶瓷經濟史, 330–1, 348, 456–7.

13 Wang and Yin, 現代景德鎮陶瓷經濟史, 331; Jingdezhen City Records Compilation Committee, 景德鎮市志略, 45, 50–1; Ding, 景德镇卷, 172.

14 Wang and Yin, 現代景德鎮陶瓷經濟史, 459.

15 Wang and Yin, 現代景德鎮陶瓷經濟史, 313, 350, 352; Jingdezhen City Records Compilation Committee, 景德鎮市志略, 329; see also Ding, 景德镇卷, 241–2.

16 Jingdezhen City Records Compilation Committee, 景德鎮市志略, 324.

17 Jingdezhen City Records Compilation Committee, 景德鎮市志略, 324; Wang and Yin, 現代景德鎮陶瓷經濟史, 331.

18 Wang and Yin, 現代景德鎮陶瓷經濟史, 318; Jingdezhen City Records Compilation Committee, 景德鎮市志略, 324.

19 Jingdezhen City Records Compilation Committee, 景德鎮市志略, 329.

20 While this initiative is not recorded in government documents or other Chinese-language histories of Jingdezhen, I met five ceramists who were part of the group. My description is based on interviews with them. One reported that a total of 200 young people had been chosen to participate.

21 Wang and Yin, 現代景德鎮陶瓷經濟史, 323, 329, 392.

22 Wang and Yin, 現代景德鎮陶瓷經濟史, 393.

23 See Andrew Walder, 'Social Change in Post-Revolution China', *Annual Review of Sociology* 15 (1989), 406–9.

24 See Qian Yingyi, 'How Reform Worked in China', *In Search of Prosperity: Analytic Narratives on Economic Growth*, ed. Dani Rodrik (Princeton: Princeton University Press, 2003), 306–7, 310, for a discussion of dual-track pricing in the early reform period. See Ding, 景德镇卷, 246, for a description of dual-track pricing in Jingdezhen's porcelain industry.

25 Wang and Yin, 現代景德鎮陶瓷經濟史, 390.

26 Ibid., 409.

27 Wang and Yin, 現代景德鎮陶瓷經濟史, 393; Jingdezhen City Records Compilation Committee, 景德鎮市志略, 50.

28 Wang and Yin, 現代景德鎮陶瓷經濟史, 416.

29 Wang and Yin, 現代景德鎮陶瓷經濟史, 414; Jingdezhen City Records Compilation Committee, 景德鎮市志略, 331.

30 Wang and Yin, 現代景德鎮陶瓷經濟史, 456; Zhou, 景德鎮史話, 1.

31 Wang and Yin, 現代景德鎮陶瓷經濟史, 456; Jingdezhen City Records Compilation Committee, 景德鎮市志略, 331.

32 Jingdezhen City Records Compilation Committee, 景德鎮市志略, 331.

33 Wang and Yin, 現代景德鎮陶瓷經濟史, 379.

34 Ibid., 378.

35 Ding, 景德镇卷, 246.

36 Jingdezhen City Records Compilation Committee, 景德鎮市志略, 54.

37 These figures come from the *Jingdezhen Statistical Yearbook* for 1986 (景德鎮統計年藍 1986), ed. Jingdezhen City Statistical Bureau (Jiangxi: Nationalized Jiangxi Yichun Information Publishing Factory, 1987).

38 Wang and Yin, 現代景德鎮陶瓷經濟史, 441.

39 Ibid., 359.

40 Zhou, 景德鎮史話, 187.

41 Excluding the column 'Employed Population', the figures in plain text are from the 1986 Jingdezhen Yearbook. I produced the column 'Employed Population' by adding the figures provided for workers (work units) and private sector employees. The figures in italics are from the 1999 Jingdezhen Yearbook. The figures in bold are taken from individual annual yearbooks for the years listed. I created the figures in bold and italics by adding up individual annual yearbook figures for state enterprise and collective workers (excluding village workers).

42 Jingdezhen City Records Compilation Committee, 景德鎮市志略, 48; Jiangxi Province Gazetteer Editorial Commission, 江西省陶瓷工業志, 22–3.

43 Jingdezhen City Records Compilation Committee, 景德鎮市志略, 332–3.

44 Wang and Yin, 現代景德鎮陶瓷經濟史, 453.

45 Jingdezhen City Records Compilation Committee, 景德鎮市志略, 54–5.

46 Jingdezhen City Statistical Bureau, 景德鎮統計年藍 1986, 3.

47 Zhou, 景德鎮史話, 186; Wang and Yin, 現代景德鎮陶瓷經濟史, 367, 411; Ding, 景德镇卷, 528.

48 Feng Shangsong was no longer head of the Labour Bureau at this time; he had moved up to a position in the Jiangxi Porcelain Company.

49 Table source: Fang Lili, 传统与变迁：景德镇旧民窑业田野调查, 43.

50 He bought the propane in canisters.

51 I interviewed Xiang Yuanhua in 2005 at the Hua Hong Porcelain Factory. Reporters for China Radio International published a slightly different version of Xiang's and Hua Hong's history in 2008. See 'Porcelain Master, Xiang Yuanhua', *CRIEnglish.com*, 29 September 2008.

52 Qian, 'How Reform Worked in China', 315; Carsten Holz, 'The Unbalanced Growth Hypothesis and the Role of the State: The Case of China's State-Owned Enterprises', *Journal of Development Economics* 96(2011), 221; see also Chen Duanjie, 'China's State-Owned Enterprises: How Much Do We Know? From CNOOC to Its Siblings', *University of Calgary The School of Public Policy Research Papers* 6(19) (2013), 22.

53 Qian, 'How Reform Worked in China', 310, 312.

54 Wang and Yin, 現代景德鎮陶瓷經濟史, 427, 430; Jian Fa, 'The Glory of China: The Nation's Staggering Porcelain Capital Epitomizes the Transition from a State Planned Economy. But It Seeks Salvation', *Beijing Weekly* 47(38) (2004), 40–3.

55 Chen Zhenghao, 'Big Moves in China City: Chaozhou Assumes the Porcelain Mantle', *Asian Ceramics* 5 (2005), 19–27.

56 Kathy Chen, 'Foshan Blends Capitalism and Socialism: China City's Market Savvy and Domestic Sales Fuel Civic Projects', *Wall Street Journal Asia*, 28 September 1992, n.p.

57 Wang Qian, 'From Local Ceramics to High-Tech Goods Exporter', *China Daily*, 9 December 2011; Wang Qian, 'Expo Revitalizes Age-Old Ceramics Industry', *China Daily*, 22 August 2012; Dai Yan, 'At Leading Edge of New Materials', *China Daily*, 27 August 2010.

58 Wang and Yin, 現代景德鎮陶瓷經濟史, 363–6.

59 Ibid., 425.

60 Ding, 景德镇卷, 241.

61 Wang and Yin, 現代景德鎮陶瓷經濟史, 469.

62 Ibid., 382.

63 Ibid., 382.

64 Ibid., 377.

65 Qian, 'How Reform Worked in China', 327; Chen, 'China's State-Owned Enterprises', 22.

66 Wang Jifu, *Strategic Challenge, Strategic Responses, and Strategies: Study of Chinese State Owned Enterprises*. (Auburn, AL: Ph.D thesis, Auburn University, 2001), 118–44; Gao Yuning, *China as the Workshop of the World* (London: Routledge Press, 2012), 144.

67 Wang and Yin, 現代景德鎮陶瓷經濟史, 405. The factory in question was called the Jingdezhen Porcelain Factory. This was not the Jingdezhen Porcelain Factory that had been diverted to making aeroplanes and cars during the Cultural Revolution; that factory was never returned to porcelain production. This Jingdezhen Porcelain Factory had been a branch of the For the People Porcelain Factory (called Factory 4369). The government separated it from For the People in 1989, and put it under the management of the Light Industry Bureau.

68 Wang and Yin, 現代景德鎮陶瓷經濟史, 384.

69 Wang and Yin, 現代景德鎮陶瓷經濟史, 377–8; Ding, 景德镇卷, 148; Chen Changgeng, Feng Shiyang and Wu Rensheng, eds., 景德鎮年鉴 1991–1992 [*Jingdezhen Yearbook 1991–1992*] (Beijing: Chinese Communist Party Central Party School Press, 1993), 182.

70 Chen et al., 景德鎮年鉴 1991–1992, 66, 214; Ding, 景德镇卷, 174–5.

71 Chen et al., 景德鎮年鉴 1991–1992, 215.

72 Chen et al., 景德鎮年鉴 1991–1992, 214–15; Wang and Yin, 現代景德鎮陶瓷經濟史, 416–17.

73 Ding, 景德镇卷, 175–6; Wang and Yin, 現代景德鎮陶瓷經濟史, 413, 439–40.

74 Jingdezhen City Records Compilation Committee, 景德鎮市志略, 49.

75 Wang and Yin, 現代景德鎮陶瓷經濟史, 439–40.
76 Chen et al., 景德鎮年鉴 1991–1992, 215.
77 Wang and Yin, 現代景德鎮陶瓷經濟史, 452.
78 Jingdezhen City Records Compilation Committee, 景德鎮市志略, 35, 41; Wang and Yin, 現代景德鎮陶瓷經濟史, 417, 436–8, 441.
79 Jingdezhen City Records Compilation Committee, 景德鎮市志略, 221.
80 Wang and Yin, 現代景德鎮陶瓷經濟史, 376–7.
81 Ibid., 399, 401, 452.
82 Wang and Yin, 現代景德鎮陶瓷經濟史, 463; Chen et al., 景德鎮年鉴 1991–1992, 371–2.
83 Ding, 景德镇卷, 292.
84 Ibid., 67.
85 Wang and Yin, 現代景德鎮陶瓷經濟史, 396, 402.
86 Chen et al., 景德鎮年鉴 1991–1992, 211.
87 Ding, 景德镇卷, 264.
88 As of the end of 1993. Ibid., 151–2.
89 Wang and Yin, 現代景德鎮陶瓷經濟史, 448.
90 Ibid., 381.

## 6 Porcelain Capital No More, 1994–2010

1 Xiong Xuehui, '景德镇十大瓷厂覆没' ['Jingdezhen's Ten Big Porcelain Factories Go Under,' 中國經營報 [*China Business News*], 18 May 2012, n.p.; Fang, 传统与变迁：景德镇旧民窑业田野调查, 83. Laid-off workers in Jingdezhen also supplied me with these dates, and the dates below. I cite printed sources for factory closures wherever available.
2 Ding, 景德镇卷, 527. The following information about Radiant Porcelain Factory comes from Ding's Appendix 4, 527–31.
3 Xiong Xuehui, '景德镇十大瓷厂覆没', n.p.
4 Patrick Tyler, 'With Deng's Influence Waning, Privatizing of China's State Industries Stalls,' *The New York Times*, 18 June 1995; Carsten Holz, 'Long Live China's State-Owned Enterprises: Deflating the Myth of Poor Financial Performance,' *Journal of Asian Economics* 13(4) (2002), 519–20.
5 Mark Frazier, 'State-Sector Shrinkage and Workforce Reduction in China,' *European Journal of Political Economy* 22(2006), 435.
6 Government records from Jingdezhen Yearbooks and Jingdezhen Statistical Yearbooks suggest a figure of about 80,000. I heard figures at the lower and higher ends of the spectrum from Jingdezhen residents. Xiong Xuehui uses both the figures of 60,000 and 100,000 in his 2012 article '景德镇十大瓷厂覆没'.

7 SOE is an acronym for state-owned enterprise. Figures in plain text are taken from individual annual yearbooks for the year listed. Figures in italics are from the 1999 Jingdezhen Yearbook. I created the figures in bold by adding up individual annual yearbook figures for state enterprise and collective workers (village workers excluded). If you compare the figures for work unit employees in 1993, just prior to the bankruptcies, and 1999, the year after the bankruptcies had concluded, you see a loss of 83,100 jobs.

8 Xiong Xuehui, '景德镇十大瓷厂覆没', n.p. I heard similar comments during my field research between 2003 and 2010.

9 Patrick Tyler, 'With Deng's Influence Waning, Privatizing of China's State Industries Stalls', n.p.

10 Carsten Holz, 'The Unbalanced Growth Hypothesis and the Role of the State: The Case of China's State-Owned Enterprises', 234; Chen Duanjie, 'China's State-Owned Enterprises', 22.

11 Frazier, 'State-Sector Shrinkage and Workforce Reduction in China', 438, 447, 449; see also Holz, 'The Unbalanced Growth Hypothesis and the Role of the State: The Case of China's State-Owned Enterprises', 221; Michael Zhang, 'The Social Marginalization of Workers in China's State-Owned Enterprises', *Social Research* 73(1) (2006), 159; see also Holz, 'Long Live China's State-Owned Enterprises: Deflating the Myth of Poor Financial Performance, 515–19'.

12 Jonathan Woetzel cites a figure of 3,658 bankruptcies. See Jonathan Woetzel, 'Reassessing China's State-Owned Enterprises', *Forbes*, 8 July 2008. See the figures for laid-off workers quoted from Frazier, above.

13 I provide several examples in 'Globalization and Deindustrialization in China's (Former) Porcelain Capital'. See also Alexandra Feytis, 'Chinese Ceramics Prepared for Further Growth', *Industrial Minerals* 526(2011), 26.

14 Chen Zhenghao, 'Big Moves in China City: Chaozhou Assumes the Porcelain Mantle', 19; Na Lan, 'Jingdezhen, Still China's Porcelain Capital?' *CRI News*, 7 August 2004, n.p.; Fully Lin and Staff Reporter, 'China's Porcelain Capital Losing Its Glory', *Want China Times*, 27 April 2012. See also Jian, 'The Glory of China'.

15 Meng Jing, 'At the Crossroads,' *China Daily*, 16 December 2011, n.p.

16 See Wayne Morrison, 'China's Economic Rise', *Congressional Research Service Report for Congress* 7–7500, 5 September 2013, 10; David Sims, 'China Widens Lead as World's Largest Manufacturer', *Thomasnet News*, 14 March 2013, n.p.

17 Wang and Yin, 現代景德鎮陶瓷經濟史, 384–5.

18 Fang, 传统与变迁：景德镇旧民窑业田野调查, 83.

19 Wang and Yin, 現代景德鎮陶瓷經濟史, 405.

20 Ibid. For information about what the China Wind Porcelain Factory produced prior to 1993, see Jingdezhen City Records Compilation Committee, 景德鎮市志略, 56.

21 Wang and Yin, 現代景德鎮陶瓷經濟史, 409.

22 Ibid., 449.

23 I describe the global crisis in ceramics manufacturing in 'Globalization and Deindustrialization in China's (Former) Porcelain Capital'. See page 220. Kathy Niblett's *Dynamic Design: The British Pottery Industry 1940–1990* (Stoke-on-Trent: City Museum and Art Gallery, 1990) has a detailed section on mergers, sellouts and takeovers in Stoke-on-Trent's ceramic industry during the 1980s. See pages 37–87.

24 Wang and Yin, 現代景德鎮陶瓷經濟史, 443; see also Ding, 景德镇卷, 180; see also Fang, 传统与变迁：景德镇旧民窑业田野调查, 46, 136.

25 Cao Min, 'Jingdezhen Reforming Enterprises', *China Business Weekly*, 2 March 1997.

26 Xiao Gongduo and Guo Guo'an report 700 private porcelain enterprises in 1993. See Xiao Gongduo and Guo Guo'an, 'Jingdezhen Rising: The Return of a Porcelain Major', *Asian Ceramics* 2 (2008), 18.

27 Fang, 传统与变迁：景德镇旧民窑业田野调查, 136.

28 Jingdezhen's sharp competition has been noted repeatedly by observers and analysts. See, for example, Imkamp, 'The Future of China: Jingdezhen Struggles to Maintain its Porcelain Fame', n.p.; Li Songjie, Wang Shujing and Li Xinghua, 'The Development of Jingdezhen in the View of Cultural Innovation', *Studies in the Sociology of Science* 3(4) (2012), 47–9; Xiao and Guo, 'Jingdezhen Rising: The Return of a Porcelain Major', 18, 22, 26; Yang Wanli, 'Porcelain Capital Hopes to Rekindle Age of Beauty', *China Daily*, 13 October 2010, n.p.

29 I met Liu Yuanchang during my fieldwork in 2004. The following description is taken from my interview with Liu Yuanchang's daughter on 20 September 2005. The interview took place in her father's studio. Biographical details about Liu's career can be found in the Chinese web-based encyclopedia *Baidu Baike* (百度百科), http://baike.baidu.com/view/896601.htm.

30 Xiong, '景德镇十大瓷厂覆没', n.p.

31 For a more detailed description of Jingdezhen's art porcelain entrepreneurs and the art porcelain market, see Maris Gillette, 'Copying, Counterfeiting, and Capitalism in Jingdezhen's Porcelain Industry', 367–403.

32 Fang, 传统与变迁：景德镇旧民窑业田野调查, 141.

33 Ibid., 176. I use the male pronoun because throwing was an exclusively male activity in Jingdezhen, and remained so through 2010 – with the exception of foreign ceramists and college-educated visitors to Jingdezhen. The wages for painters are found on Fang, 传统与变迁：景德镇旧民窑业田野调查, 180.

34 Ding, 景德镇卷, 67.

35 Xiao and Guo, 'Jingdezhen Rising: The Return of a Porcelain Major', 18, 22; see also Fang, 传统与变迁：景德镇旧民窑业田野调查, 56, 197.

36 Yang, 'Porcelain Capital Hopes to Rekindle Age of Beauty', n.p.

37 Fang, 传统与变迁：景德镇旧民窑业田野调查, 88.

38 I describe these workshops in detail in 'Copying, Counterfeiting and Capitalism in Jingdezhen's Porcelain Industry'.

39 Jonathan Jones, 'Scandal in China over the Museum with 40,000 Fake Artefacts,' *The Guardian*, 17 July 2013. www.theguardian.com/culture/2013/jul/17/jibaozhai-museum-closed-fakes-china.

40 As described in Gillette, 'Copying, Counterfeiting, and Capitalism in Jingdezhen's Porcelain Industry'.

41 McKirdy, '"Holy Grail" of Chinese Porcelain Nets Record Bid for Sotheby's', n.p.

42 Frederik Balfour, 'The Expensive Antics of China's Gaudiest Billionaire', *Bloomberg Business*, 16 April 2015, n.p.

43 Jingdezhen City Gazetteer Editorial Committee, 景德镇年蓝, 223.

44 Ding, 景德镇卷, 188–93.

45 Fang, 传统与变迁：景德镇旧民窑业田野调查, 89; see also Li et al., 'The Development of Jingdezhen in the View of Cultural Innovation', 49.

46 Many observers note the prevalence of copying and counterfeiting in Jingdezhen. See for example Gillette, 'Copying, Counterfeiting, and Capitalism in Jingdezhen's Porcelain Industry'; Li et al., 'The Development of Jingdezhen in the View of Cultural Innovation', 47–9; Imkamp, 'The Future of China: Jingdezhen Struggles to Maintain its Porcelain Fame', n.p.; Daphne Lange Rosenzweig, 'Continuity and Creativity: Modern Jingdezhen Ceramics', *International Chinese Snuff Bottle Society Journal* 35 (Autumn 2003), 29, and so on.

47 Yang, 'Porcelain Capital Hopes to Rekindle Age of Beauty', n.p. A senior named Gao Ling reported a commission for perfume bottles and told the reporter that he had a classmate who earned 'tens of thousands of yuan a year through his workshop business'. I met several Jingdezhen Ceramics Institute students who produced commercial porcelain art.

48 I discuss this in detail in 'Copying, Counterfeiting, and Capitalism in Jingdezhen's Porcelain Industry'. Participants in my film *Broken Pots Broken Dreams* also describe some of the difficulties of making money at art porcelain production.

49 Xiao and Guo, 'Jingdezhen Rising: The Return of a Porcelain Major', 20.

50 Details about Franz Porcelain and its founder come from the company's website, www.franzcollection.com. *Elle* magazine published a short piece on Franz Porcelain in the 'Elle Décor – Steeped in Style' section of their online publication, which I accessed on 9 October 2013. See Laura Regensdorf, 'Steeped in Style: Jean Boggio for Franz's Porcelain Tea Service', *Elle Décor*, n.p.

51 Franz Collection Inc., 'Franz Collection Inc. President Brings Together Porcelain Manufacturers From All Over The World', Press Release, 7 July 2004. www.franzcollection.com/news/press_release.php?press_release_id=6.

52 As described on the company's website. See 'Franz Collection Inc. Goes Green', www.franzcollection.com/about/going_green.php.

53 Xiao and Guo, 'Jingdezhen Rising', 18, 24, 26.

54 Ibid., 24.

55 Li et al., 'The Development of Jingdezhen in the View of Cultural Innovation', 47–9.

56 Xiong, '景德镇十大瓷厂覆没', n.p. The analyst mentioned in the following sentence is Liu Yuanping, of the Chinese Academy of Social Sciences.

## 7 From Porcelain Capital to Heritage Site

1 Hanqin Qiu Zhang, Ray Pine and Terry Lam, *Tourism and Hotel Development in China: From Political to Economic Success* (Binghamton, NY: Haworth Hospitality Press, 2005), 15.

2 Zhang et al., *Tourism and Hotel Development in China: From Political to Economic Success*, 48, 51, 102–7; Travis Klingberg and Tim Oakes, 'Producing Exemplary Consumers: Tourism and Leisure Culture in China's Nation-Building Project', *China In and Beyond the Headlines*, eds. Timothy B. Weston and Lionel M. Jensen (Lanham, MD: Rowman & Littlefield, 2012), 196–7, 200–1.

3 Klingberg and Oakes, 'Producing Exemplary Consumers: Tourism and Leisure Culture in China's Nation-Building Project', 204.

4 Vincent Wolfington and Mark Wolfington, 'The Hospitality Talent Gap', *China Business Review*, 1 January 2012.

5 Hong Kong Trade Development Research, 'Jingdezhen (Jiangxi) City Information', 14 January 2014, 2. See http://china-trade-research.hktdc.com/business-news/article/Fast-Facts/Jingdezhen-Jiangxi-City-Information/ff/en/1/1X000000/1X09VZCR.htm.

6 For example, Zoe Li, 'Interview: Caroline Cheng on Tea, Clay, and China', Blouin Artinfo, Hong Kong, http://hk.blouinartinfo.com/news/story/976443/interview-caroline-cheng-on-tea-clay-and-china#; Violet Law, 'Intelligent Design: Ceramic Artists From All Over the Mainland are Descending on Jingdezhen as the Jiangxi City, Famed for Its Pottery Tradition, Enjoys a Creative Renaissance', *South China Morning Post*, 29 September 2013, 64–5; Huan Hsu, 'Five Reasons to Visit Jingdezhen', *Fodor's Travel*, 6 April 2015, www.fodors.com/news/5-reasons-to-visit-jingdezhen-11268.html.

7 Sources for this table are the Deputy Party Secretary from Jingdezhen's Cultural Affairs Bureau, Mimi Li and and Wu Bihu, *Urban Tourism in China* (London: Routledge, 2012), 19; and Jingdezhen City Statistical Office and National Statistics Office Jingdezhen Investigation Team, '景德镇市 2014 年国民经济和社会发展统计报告' [Report on the Economic and Social Development of Citizens in Jingdezhen City, 2014], 景德镇在线, 13 April 2015, www.jdzol.com/2015/0413/113428.html.

8 Philip Tinari, 'Postures in Clay: The Vessels of Ai Weiwei', *Ai Weiwei: Dropping the Urn, Ceramic Works, 5000 BCE to 2010 CE* (Glenside, PA: Arcadia University Art

Gallery, 2010), 36; Jane Chin Davidson, 'Affirmative Precarity: Ai Weiwei and Margarita Cabrera', *Journal of Visual Culture* 12(128) (2013): 136. The figure of 100 million sunflower seeds is from 'Ai Weiwei Sunflower Seeds', a website created by the Faurschou Foundation in Denmark. See www.aiweiweiseeds.com/about-ai-weiweis-sunflower-seeds.

9 Tinari, 'Postures in Clay: The Vessels of Ai Weiwei', 40.

10 See Brankston, *Early Ming Wares at Jingdezhen*, 56.

11 Karen Terpstra and Zhu Gui Hong, 'Firing Methods of a Wood-fired Jingdezhen Commercial Kiln', *Ceramics Technical* 12 (May 2001): 25–9.

12 In November 1959, the Jiangxi Provincial Government declared three kiln sites, one of which was Hutian, conservation areas. Work units were prohibited from engaging in any activities that would damage the sites. See Jingdezhen City Records Compilation Committee, 景德鎮市志略, 314.

13 Zhou, 景德鎮史話, 186; Jingdezhen City Records Compilation Committee, 景德鎮市志略, 329.

14 Jingdezhen City Gazetteer Editorial Group, 景德镇年鉴 2003 [*Jingdezhen Yearbook 2003*] (Jiangxi: Jiangxi Dizhi Cehui Printing Factory, 2004), 266.

15 Wang and Yin, 現代景德鎮陶瓷經濟史, 446.

16 *PR Newswire*, 'Jingdezhen Holds Kiln Reactivation Ceremony for Ming and Qing Royal Kilns: The Blue and White, Dragon Pattern Vat and Wind and Fire Kilns', 30 October 2013, *PR Newswire*, www.prnewswire.com/news-releases/jingdezhen-holds-kiln-reactivation-ceremony-for-ming-and-qing-dynasty-royal-kilns-the-blue-and-white-dragon-pattern-vat-and-wind-and-fire-kilns-229885601.html#.

17 Paul Harris, 'Porcelain Orchestra Symbolizes Achievements of Jingdezhen', Chinese Art Blog: Monitoring the UK Marketplace, 5 April 2015, http://chineseart.co.uk/tag/jingdezhen.

18 Jingdezhen City Government, 'Ceramic Schools and Institutes', *China.Jingdezhen*. http://eng.jdz.gov.cn/Online/Ceramic/201204/t20120405_133150.htm, 2011b.

19 See 'Important Discoveries in Jingdezhen', *Xinhua* [New China] News Agency, 28 January 2004; and 'Imperial Kilns in Zhushan of Jingdezhen, Jiangxi Province', 12 April 2004, China.org.cn.

20 Alfred University Press Release, 'Shanghai University Honors AU Artist', 1 December 2000, www.alfred.edu/pressreleases/viewrelease.cfm?ID=449#sthash.DWUad8Ki.dpuf.

21 Peng Yining, 'Preserving the Ancient Traditions', *China Watch*, 21 July 2014.

22 See Alfred University Press Release, 'Shanghai University Honors AU Artist'. Sixteen Alfred University students attended workshops at Sanbao in the summer of 2000.

23 Peng writes in 'Preserving the Ancient Traditions' that Li purchased four hammers in 2013, at a cost of $27,300 USD, to prevent them from being torn down.

24 For examples of publications, see Yonina Blech-Hermioni, 'A Summer Residency in China', *Ceramics Monthly* 50(1) (January 2002), 74–7; Stephen Brosseau, 'Throwing Classical Porcelain in Jingdezhen, China', *Ceramics Technical* 18 (2004), 45–8; Colin Martin, 'International Ceramic Artists Working in China', *Craft Arts International* 86 (2012), 97–9. See also Violet Law, 'Intelligent Design', 65; and Anna Wu, 'Jingdezhen Today', *Arts of Asia* 41(6) (2011), 94.

25 You can see aspects of Lislerud's process in a short video from 2013 by Dennis Appelong entitled 'The Wall Is In Your Mind', available on Vimeo at https://vimeo.com/62236662.

26 For a characterization of Lislerud's work, see Rob Barnard et al.'s *Breaking the Mould: New Approaches to Ceramics* (London: Black Dog Publishing, 2007), 198–9.

27 Quote from Robyn Phelan, '15 Days in Jingdezhen,' *Ceramics Technical* 30 (2010), 75. For information about the Saturday Creative Market see Li, 'Interview: Caroline Cheng on Tea, Clay, and China', n.p.

28 Phelan, '15 Days in Jingdezhen', 76.

29 Ibid., 77; see also 75–6.

30 Ken Weiderman, 'Porcelain Reborn: CR's Group Show Honors the Porcelain City', *The North Coast Journal*, 5 March 2015. www.northcoastjournal.com/humboldt/porcelain-reborn/Content?oid=2835844.

31 *Global Times*, 'Heart of China', 21 July 2015, www.globaltimes.cn/content/933114.shtml.

32 Anna Wu discusses four international artists who used local labour in her 2011 article 'Jingdezhen Today'; see page 94. Colin Martin describes ten such foreign artists in his 2012 article 'International Artists Working in China'.

33 Wu, 'Jingdezhen Today', 96.

34 Phelan, '15 Days in Jingdezhen', 78.

35 *Time Out Shanghai*, 'Weekend Break: Jingdezhen', 16 November 2012, n.p., www.timeoutshanghai.com/features/Travel-Weekend_breaks/8873/Weekend-break-Jingdezhen.html.

36 Gert-Jen Hospers, 'Industrial Heritage Tourism and Regional Restructuring in the European Union', *European Planning Studies* 10(3) (2002); Tiffany Jenkins, 'Can Culture Save Cities?' *BBC*, 21 October 2014, n.p.; Charles Landry et al., *The Art of Regeneration: Urban Renewal Through Cultural Activity* (Gloucester: Commedia, 1996), 32.

37 On 27 July 2015 I located seven tour companies that were offering or had recently offered such tours: Huangshan CITS, China Connection Tours, Martin Randall Travel, Chinese Clay Art Tours, Diverse China, China Links Travel, and Quadrant Travel and Journeys.

38 Wang Jianguo and Jiang Nan, 'Conservation and Adaptive-Reuse of Historical Industrial Building in China in the Post-Industrial Era', *Frontiers of Architecture and Civil Engineering in China* 1(4) (2007), 474–80.

39 See Law, 'Intelligent Design', 64; Martin, 'International Ceramic Artists Working in China', 99; and the Pottery Workshop website 'About Us', www.potteryworkshop.com.cn/aboutus.asp.

40 *International Business Times*, 'Chinese Dissident Ai Weiwei's Sunflower Seeds Pull In $782,000', 11 May 2012, n.p.

41 Ancient Kiln Folk Custom Museum Press Release, 'Jingdezhen Applies for Place on World Heritage List', 11 July 2013, www.multivu.com/mnr/62383-jingdezhen-ancient-kiln-applies-for-place-on-world-heritage-list.

# Bibliography


Alfred University Press Release. 2000. 'Shanghai University Honors AU Artist'. 1 December 2000. www.alfred.edu/pressreleases/viewrelease.cfm?ID=449 sthash.DWUad8Ki.dpuf.

Ancient Kiln Folk Custom Museum Press Release. 2013. 'Jingdezhen Applies for Place on World Heritage List.' *Ancient Kiln Folk Custom Museum* website, 11 July 2013. www.multivu.com/mnr/62383-jingdezhen-ancient-kiln-applies-for-place-on-world-heritage-list.

Appelong, Dennis. 2013. 'The Wall Is In Your Mind.' n.p. [film: 2'46']. https://vimeo.com/62236662.

*Baidu Baike*. '刘远长' ['Liu Yuanchang']. 百度百科 [*Baidu Baike*]. http://baikc.baidu.com/view/896601.htm.

Balfour, Frederik. 2015. 'The Expensive Antics of China's Gaudiest Billionaire'. *Bloomberg Business*, 16 April 2015. www.bloomberg.com/news/articles/2015-04-17/the-expensive-antics-of-china-s-gaudiest-billionaire.

Barnard, Rob, Natasha Daintry and Clare Twomey. 2007. *Breaking the Mould: New Approaches to Ceramics*. London: Black Dog Publishing.

Bastid-Bruguiere, Marianne. 1980. 'Currents of Social Change'. In *Cambridge History of China: Late Ch'ing*, volume 11, part 2, eds. John K. Fairbank and Kwang-Ching Liu. Cambridge: Cambridge University Press, 535–602.

Blech-Hermioni, Yonina. 2002. 'A Summer Residency in China.' *Ceramics Monthly* 50(1): 74–7.

Brankston, Archibald Dooley. 1938. *Early Ming Wares at Jingdezhen*. Peking: H. Vetch.

Brosseau, Stephen. 2004. 'Throwing Classical Porcelain in Jingdezhen, China'. *Ceramics Technical* 18: 45–8.

Cao Min. 1997. 'Jingdezhen Reforming Enterprises'. *China Business Weekly*, 2 March 1997. www.chinadaily.com.cn/epaper/html/bw/1997/199703/19970302/19970302005_2.html.

Carswell, John. 2000. *Blue and White: Chinese Porcelain Around the World*. Chicago, IL: Art Media Resources.

Chaffee, John W. 1999. *Branches of Heaven: A History of the Imperial Clan of Sung China*. Cambridge, MA: Harvard University Press.

Chen Changgeng, Feng Shiyang and Wu Rensheng, eds. 1993. 景德鎮年鉴 1991–1992 [*Jingdezhen Yearbook 1991–1992*]. Beijing: Chinese Communist Party Central Party School Press.

Chen Duanjie. 2013. 'China's State-Owned Enterprises: How Much Do We Know? From CNOOC to Its Siblings'. *University of Calgary The School of Public Policy Research Papers* 6(19): 1–27.

Chen, Kathy. 1992. 'Foshan Blends Capitalism and Socialism: China City's Market Savvy and Domestic Sales Fuel Civic Projects.' *Wall Street Journal Asia*, 28 September 1992.

Chen Yuqian, ed. 2004. 景德鎮陶瓷文化概論 [*A Discussion of Jingdezhen's Ceramic Culture*]. Nanchang, Jiangxi: Jiangxi Higher Education Publishers.

Chen Zhenghao. 2005. 'Big Moves in China City: Chaozhou Assumes the Porcelain Mantle'. *Asian Ceramics* 5: 19–27.

China.org.cn. 2004. 'Imperial Kilns in Zhushan of Jingdezhen, Jiangxi Province'. *Top Ten Archaeological Discoveries of 2003,* 12 April 2004. www.china.org.cn/english/features/Archaeology/98175.htm#10.

*China Radio International*. 2008. 'Porcelain Master, Xiang Yuanhua'. *CRIEnglish.com*. english.cri.cn/4406/2008/09/29/1141s410143.htm.

Clennell, W.J. 1905 [1983]. 'A Report on a Journey in the Interior of Kiangsi'. In *Ching-Te-Chen: Views of a Porcelain City*, ed. Robert Tichane. Painted Post, NY: New York State Institute for Glaze Research, 315–36.

Clunas, Craig. 1991. 'The Cost of Ceramics and the Cost of Collecting Ceramics in the Ming Period'. *Bulletin of the Oriental Ceramics Society of Hong Kong* 8 (1986–1988): 47–52.

Cosentino, Frank. 1976. *The Boehm Journey to Ching-te-Chen, China, Birthplace of Porcelain*. Trenton, NJ: Edward Marshall Boehm, Inc.

Dai Yan. 2010. 'At the Leading Edge of New Materials'. *China Daily,* 27 August 2010. www.chinadaily.com.cn/cndy/2010-08/27/content_11211108.htm.

Davidson, Jane Chin. 2013. 'Affirmative Precarity: Ai Weiwei and Margarita Cabrera'. *Journal of Visual Culture* 12(128): 128–47.

D'Entrecolles, Francois. 1712, 1722 [1983]. Letter I and II. In *Ching-Te-Chen: Views of a Porcelain City*, ed. Robert Tichane. Painted Post, NY: New York State Institute for Glaze Research, 51–128.

Dikötter, Frank. 2010. *Mao's Great Famine: The History of China's Most Devastating Catastrophe, 1958–1962*. New York: Walker and Company.

Dillon, Michael. 1976. *A History of the Porcelain Industry in Jingdezhen*. Ph.D thesis, University of Leeds.

Dillon, Michael. 1992. 'Fang Zhimin, Jingdezhen, and the Northeast Jiangxi Soviet: Tradition, Revolution, and Civil War in a Pottery Town'. *Modern Asian Studies* 26(3): 569–89.

Dillon, Michael. 2007. *Key Papers On Chinese Economic History Up To 1949*. Folkestone: Global Oriental.

Ding Weizhi, ed. 1996. 景德镇卷 [*A Record of Jingdezhen*]. Beijing: China Encyclopedia Publisher.

Fang Lili. 2000. 传统与变迁：景德镇旧民窑业田野调查 [*Traditions and Changes: Investigations into Jingdezhen's Old Folk Potteries*]. Nanchang: Jiangxi People's Publisher.

Fang Lili. 2002. 景德鎮民窰 [*Jingdezhen's Folk Potteries*]. Beijing: Renmin meishu chubanshe.

Faurschou Foundation. n.d. 'Ai Weiwei Sunflower Seeds'. www.aiweiweiseeds.com/about-ai-weiweis-sunflower-seeds.

Feng Shangsong. 2008. 我的七十年回顾 [*My Memories at 70*]. Jingdezhen: n.p.

Feuerwerker, Albert. 1980. 'Economic Trends in the Late Ch'ing Empire, 1870–1911'. In *Cambridge History of China: Late Ch'ing*, volume 11, part 2, eds. John K. Fairbank and Kwang-Ching Liu. Cambridge: Cambridge University Press, 1–69.

Feytis, Alexandra. 2011. 'Chinese Ceramics Prepared for Further Growth'. *Industrial Minerals* 526 (July): 26–7.

Finlay, Robert. 1998. 'The Pilgrim Art: The Culture of Porcelain in World History'. *Journal of World History* 9(2): 141–87.

Franck, Harry. 1925. *Roving Through Southern China*. New York: The Century Company.

Franz Collection, Inc. n.d. 'The Creator and Innovator: Francis Chen'. Franz website. www.franzcollection.com/about/creator.php.

Franz Collection, Inc. n.d. 'Franz Collection Inc. Goes Green'. Franz website. www.franzcollection.com/about/going_green.php.

Franz Collection, Inc. 2004. 'Franz Collection Inc. President Brings Together Porcelain Manufacturers From All Over The World'. Press Release, 7 July 2004. www.franzcollection.com/news/press_release.php?press_release_id=6.

Frazier, Mark. 2006. 'State-Sector Shrinkage and Workforce Reduction in China'. *European Journal of Political Economy* 22: 435–51.

Gao Yuning. 2012. *China as the Workshop of the World*. London: Routledge Press.

Garnsey, Wanda and Rewi Alley. 1983. *China: Ancient Kilns and Modern Ceramics: A Guide to the Potteries*. Canberra: Australian National University Press.

Gerritsen, Anne. 2009. 'Fragments of a Global Past: Ceramics Manufacture in Song-Yuan-Ming Jingdezhen'. *Journal of the Economic and Social History of the Orient* 52: 117–52.

Gillette, Maris Boyd. 2009. *Broken Pots Broken Dreams* [film: 28' 41']. Haverford, PA: n.p.

Gillette, Maris Boyd. 2010. 'Copying, Counterfeiting, and Capitalism in Jingdezhen's Porcelain Industry'. *Modern China* 36: 367–403.

Gillette, Maris Boyd. 2012. 'Contemporary Jingdezhen's Porcelain Entrepreneurs'. In *Landscapes in Blue – Made in China: The Vase Project*. Lafayette, PA: Lafayette University, 44–53.

Gillette, Maris Boyd. 2014a. 'Labor and Precariousness in China's Porcelain Capital', *Anthropology of Work Review* 35: 25–39.

Gillette, Maris Boyd. 2014b. 'Documenting, Dramatizing, and Representing China's Porcelain World in *Broken Pots Broken Dreams*', *Visual Anthropology Review* 30: 38–49.

Gillette, Maris Boyd. 2015. 'Globalization and Deindustrialization in China's (Former) Porcelain Capital'. In *East Asia in the World: An Introductory Guide*, ed. Anne Prescott. Armonk, NY: M.E. Sharpe, Inc., 219–42.

*Global Times*. 2015. 'Heart of China', 21 July 2015. www.globaltimes.cn/content/933114.shtml.

*The Guardian*. 2014. 'Chinese Museum Closed Down Over "Fake" Exhibits', 22 May 2014. www.theguardian.com/world/2014/may/22/chinese-museum-closed-down-fake-exhibits-lucheng.

Harris, Paul. 2015. 'Porcelain Orchestra Symbolizes Achievements of Jingdezhen'. Chinese Art Blog: Monitoring the UK Marketplace, 5 April 2015. http://chineseart.co.uk/tag/jingdezhen.

Harrison-Hall, Jessica. 2001. 'The Ming Porcelain Industry at Jingdezhen 1368–1644'. In *Catalogue of Late Yuan and Ming Ceramics in the British Museum*. London: The British Museum Press, 19–25.

Hobson, R.L. 1915 [1976]. *Chinese Pottery and Porcelain*. New York: Dover Publications Inc.

Holz, Carsten. 2002. 'Long Live China's State-Owned Enterprises: Deflating the Myth of Poor Financial Performance.' *Journal of Asian Economics* 13(4): 493–529.

Holz, Carsten. 2011. 'The Unbalanced Growth Hypothesis and the Role of the State: The Case of China's State-Owned Enterprises'. *Journal of Development Economics* 96: 220–38.

Hong Kong Trade Development Research. 2013. 'Jingdezhen (Jiangxi) City Information'. *Hong Kong Trade Development Council*, 14 January 2014. http://china-trade-research.hktdc.com/business-news/article/Fast-Facts/Jingdezhen-Jiangxi-City-Information/ff/en/1/1X000000/1X09VZCR.htm, 1–3.

Hospers, Gert-Jen. 2002. 'Industrial Heritage Tourism and Regional Restructuring in the European Union'. *European Planning Studies* 10(3): 397–404.

Hsu, Huan. 2015. 'Five Reasons to Visit Jingdezhen'. *Fodor's Travel*, 6 April 2015. www.fodors.com/news/5-reasons-to-visit-jingdezhen-11268.html.

Imkamp, Laura. 2011. 'The Future of China: Jingdezhen Struggles to Maintain its Porcelain Fame'. *CNN Travel*, 20 July 2011, n.p. http://travel.cnn.com/shanghai/shop/future-china-jingdezhen-struggles-keep-its porcelain-fame-835545.

*International Business Times*. 2014. 'Chinese Dissident Ai Weiwei's Sunflower Seeds Pull In $782,000', 11 May 2012. www.ibtimes.com/chinese-dissident-ai-weiweis-sunflower-seeds-pull-782000–698049.

Jenkins, Tiffany. 2014. 'Can Culture Save Cities?' *BBC*, 21 October 2014. www.bbc.com/culture/story/20140502-can-culture-save-cities.

Jian Fa. 2004. 'The Glory of China: The Nation's Staggering Porcelain Capital Epitomizes the Transition from a State Planned Economy. But it Seeks Salvation'. *Beijing Weekly* 47(38): 40–3.

Jiang Qi. 1322 [1985]. 陶記 [*Ceramics Memoir*]. Translated by Stephen Bushell. In *Ching-Te-Chen: Views of a Porcelain City*, ed. Robert Tichane. Painted Post, NY: New York State Institute for Glaze Research, 43–8.

Jiang Siqing. 1936. 景德鎮瓷業史 [*A History of Jingdezhen's Porcelain Industry*]. Shanghai: Zhonghua Shuju.

Jiangxi Light Industry Bureau Ceramics Research Center. 1959. 景德鎮陶瓷史稿 [*A Draft History of Jingdezhen Ceramics*]. Beijing: Sanlian Publishers.

Jiangxi Province Gazetteer Editorial Committee, eds. 2005. 江西省陶瓷工業志 [*A Record of the Ceramics Industry in Jiangxi Province*]. Jiangxi: Jiangxi Province Gazetter Publishing.

Jingdezhen City Gazetteer Editorial Group. n.d. (1994). 景德镇年鉴 1993 [*Jingdezhen Yearbook 1993*]. n.p.: n.p.

Jingdezhen City Gazetteer Editorial Group. 1995. 景德镇年监 1994 [*Jingdezhen Yearbook 1994*]. Jiangxi: Ruifeng Caise Publishing Company Ltd.

Jingdezhen City Gazetteer Editorial Group. 1996. 景德镇年监 1995 [*Jingdezhen Yearbook 1995*]. Jiangxi: Jiangxi Dizhi Cehui Printing Factory.

Jingdezhen City Gazetteer Editorial Group. 1997. 景德镇年监 1997 [*Jingdezhen Yearbook 1997*]. Jiangxi: Jiangxi Dizhi Cehui Printing Factory.

Jingdezhen City Gazetteer Editorial Group. 1999. 景德镇年鉴 1998 [*Jingdezhen Yearbook 1998*]. Jiangxi: Jiangxi Dizhi Cehui Printing Factory.

Jingdezhen City Gazetteer Editorial Group. 2004. 景德镇年鉴 2003 [*Jingdezhen Yearbook 2003*]. Jiangxi: Jiangxi Dizhi Cehui Printing Factory.

Jingdezhen City Government. 2011a. 'A Brief Introduction of [sic] Jingdezhen'. *China. Jingdezhen*. http://eng.jdz.gov.cn/Brief/introduction/201112/t20111201_122542.htm.

Jingdezhen City Government. 2011b. 'Ceramic Schools and Institutes'. *China.Jingdezhen*. http://eng.jdz.gov.cn/Online/Ceramic/201204/t20120405_133150.htm.

Jingdezhen City Porcelain Industry Wage Reform Office. 1956. 景德鎮市瓷業工資改革資料彙集 [*Compilation of Jingdezhen City Porcelain Industry Wage Reform Data*]. n.p.: n.p.

Jingdezhen City Records Compilation Committee, eds. 1989. 景德鎮市志略 [*Brief Records of Jingdezhen City*]. Shanghai: Chinese Dictionary Publishers.

Jingdezhen City Statistical Bureau, ed. 1987. 景德镇统计年鉴 1986 [*Jingdezhen Statistical Yearbook 1986*]. Jiangxi: Nationalized Jiangxi Yichun Information Publishing Factory.

Jingdezhen City Statistical Bureau, ed. 1999. 景德镇统计年鉴 1999 [*Jingdezhen Statistical Yearbook 1999*]. Jiangxi: Nationalized Jiangxi Yichun Information Publishing Factory.

Jingdezhen City Statistical Office and National Statistics Office Jingdezhen Investigation Team. 2015. '景德镇市 2014 年国民经济和社会发展统计报告' ['Report on the Economic and Social Development of Citizens in Jingdezhen City, 2014']. 景德镇在线, 13 April 2015. www.jdzol.com/2015/0413/113428.html.

Jingdezhen Historical Ceramic Industry Research Team, Jiangxi Province History Association, Jingdezhen Research Team for the Study of Ceramics Production. 1963. 景德镇制瓷业历史调查资料选辑 [*Compiled Data from Historical Investigation Into Jingdezhen's Porcelain Industry*]. Jiangxi: n.p.

Jingdezhen Statistical Yearbook. 2003. 景德鎮統計年鑒 [*Jingdezhen Statistical Yearbook*]. Nanchang: Jiangxi dizhi cehui yinshuachang.

Jingdezhen Yearbook Gazetteer Editorial Group, ed. 2000. 景德镇年蓝 1999 [*Jingdezhen Yearbook 1999*]. Jiangxi: Jiangxi Ruifeng Caise Publishing Company, Ltd.

Jingdezhen Yearbook Gazetteer Editorial Group, ed. 2001. 景德镇年蓝 2000 [*Jingdezhen Yearbook 2000*]. Jiangxi: Jiangxi Ruifeng Caise Publishing Company, Ltd.

Jingdezhen Yearbook Gazetteer Editorial Group, ed. 2002. 景德镇年蓝 2001 [*Jingdezhen Yearbook 2001*]. Jiangxi: Jiangxi Ruifeng Caise Publishing Company, Ltd.

Jingdezhen Yearbook Gazetteer Editorial Group, ed. 2003. 景德镇年蓝 2002 [*Jingdezhen Yearbook 2002*]. Jiangxi: Jiangxi Ruifeng Caise Publishing Company, Ltd.

Jones, Jonathan. 2013. 'Scandal in China over the Museum with 40,000 Fake Artefacts'. *The Guardian*, 17 July 2013. www.theguardian.com/culture/2013/jul/17/jibaozhai-museum-closed-fakes-china.

Kerr, Rose. 1986. *Chinese Ceramics: Porcelain of the Qing dynasty 1644–1911*. London: Victoria and Albert Museum.

Kerr, Rose and Nigel Wood. 2004. *Ceramic Technology. Science and Civilization in China* 5(12), ed. Joseph Needham. Cambridge: Cambridge University Press.

Klingberg, Travis and Tim Oakes. 2012. 'Producing Exemplary Consumers: Tourism and Leisure Culture in China's Nation-Building Project'. In *China In and Beyond the Headlines*, eds. Timothy B. Weston and Lionel M. Jensen. Lanham, MD: Rowman & Littlefield, 195–213.

Landry, Charles, Lesley Greene, Francois Matarasso and Franco Bianchini. 1996. *The Art of Regeneration: Urban Renewal Through Cultural Activity*. Gloucester: Commedia.

Lardy, Nicholas R. 1987. 'Economic Recovery and the First Five Year Plan'. In *The Cambridge History of China* 14, eds. Roderick MacFarquhar and John King Fairbank. Cambridge: Cambridge University Press, 144–83.

Lardy, Nicholas R. 1987. 'The Chinese Economy Under Stress, 1958–1965'. In *The Cambridge History of China* 14, eds. Roderick MacFarquhar and John King Fairbank. Cambridge: Cambridge University Press, 360–97.

Lau, Nap-yin and Huang K'uan-chung. 2009. 'A New Type of Emperor: The Diffident Zhenzong, 997–1022'. In *The Cambridge History of China* 5(1), eds. Denis Twitchett and Paul Smith. Cambridge: Cambridge University Press, 260–78.

Law, Violet. 2013. 'Intelligent Design: Ceramic Artists From All Over the Mainland Are Descending on Jingdezhen as the Jiangxi City, Famed for Its Pottery Tradition, Enjoys a Creative Renaissance'. *South China Morning Post*, 29 September 2013, 64–5.

Lenz, Frank B. 1920. 'The World's Ancient Porcelain Center'. *The National Geographic Magazine* 38(5): 391–406.

Li, Jackson (Li Jianshen). n.d. 'Walk through Sanbao'. http://chinaclayart.com/ [film: 7'43'].

Li, Mimi and Wu Bihu. 2012. *Urban Tourism in China*. London: Routledge.

Li Songjie, Wang Shujing and Li Xinghua. 2012. 'The Development of Jingdezhen in the View of Cultural Innovation'. *Studies in the Sociology of Science* 3(4): 45–9.

Li, Zoe. 2013. 'Interview: Caroline Cheng on Tea, Clay, and China', *Blouin Artinfo*, Hong Kong, 25 October 2013. http://hk.blouinartinfo.com/news/story/976443/interview-caroline-cheng-on-tea-clay-and-china#.

Liang Miaotai. 1991. 明清景德镇城市经济研究 [*Research on Jingdezhen's Urban Economy During the Ming and Qing Dynasties*]. Nanchang: Jiangxi People's Publisher.

Lin, Fully and Staff Reporter. 2012. 'China's Porcelain Capital Losing Its Glory'. *Want China Times*, 27 April 2012. www.wantchinatimes.com/news-subclass-cnt.aspx?id=20120427000009&cid=1504.

Luo Xuezheng. 2004. '景德鎮明清官窯落選瓷的變賣與防止' ['The Sale and Reproduction of Imperial Kiln Seconds of the Ming and Qing Dynasties']. 文物天地 [*Wenwu Tiandi*] 12: 28–30.

Luo Xuezheng. 2005. '景德鎮仿古瓷的歷史與現狀' ['The History and Current Situation of Antique Reproductions in Jingdezhen']. 美術 [*Meishu*] 7(7): 100–3.

Mallory, Walter. 1926. *China: The Land of Famine*. New York: American Geographical Society.

Martin, Colin. 2012. 'International Ceramic Artists Working in China'. *Craft Arts International* 86: 97–9.

McKirdy, Euan. 2014. '"Holy Grail" of Chinese Porcelain Nets Record Bid for Sotheby's'. *CNN.com*, 10 April 2014. http://edition.cnn.com/2014/04/09/world/asia/sothebys-record-sale-chinese-porcelain/.

Meng Jing. 2011. 'At the Crossroads'. *China Daily*, 16 December 2011. http://usa.chinadaily.com.cn/weekly/2011-12/16/content_14274784.htm.

Meisner, Maurice. 1999. *Mao's China and After: A History of the People's Republic*. New York: Free Press.

Morrison, Wayne. 2013. 'China's Economic Rise'. *Congressional Research Service Report for Congress* 7-7500. 5 September 2013.

Na Lan. 2004. 'Jingdezhen: Still China's Porcelain Capital?' *CRI Online*, 7 August 2004. http://english.cri.cn/1174/2004-8-7/103@139250.htm.

Niblett, Kathy. 1990. *Dynamic Design: The British Pottery Industry 1940–1990*. Stoke-on-Trent: City Museum and Art Gallery.

Peng Yining. 2014. 'Preserving the Ancient Traditions'. *China Watch*, 21 July 2014. http://chinawatch.washingtonpost.com/2014/07/preserving_the_ancient_traditions/.

Phelan, Robyn. 2010. '15 Days in Jingdezhen'. *Ceramics Technical* 30: 74–9.

Pierson, Stacey. 2000. 'Precious Blue: Cobalt in Chinese Ceramic Glazes and Decoration'. *Taoci* 1: 33–40.

Pottery Workshop. 2010–2011. 'About Us'. *The Pottery Workshop* website. www.potteryworkshop.com.cn/aboutus.asp.

*PR Newswire*. 2013. 'Jingdezhen Holds Kiln Reactivation Ceremony for Ming and Qing Royal Kilns: The Blue and White, Dragon Pattern Vat and Wind and Fire Kilns', 30 October 2013. www.prnewswire.com/news-releases/jingdezhen-holds-kiln-reactivation-ceremony-for-ming-and-qing-dynasty-royal-kilns-the-blue-and-white-dragon-pattern-vat-and-wind-and-fire-kilns-229885601.html#.

Qian Yingyi. 2003. 'How Reform Worked in China'. In *In Search of Prosperity: Analytic Narratives on Economic Growth*, ed. Dani Rodrik. Princeton: Princeton University Press, 297–333.

Regensdorf, Laura. 2013. 'Steeped in Style: Jean Boggio for Franz's Porcelain Tea Service'. *Elle Décor*. www.elledecor.com/SHOPPING/WHAT-WE-LOVE/STEEPED-IN-STYLE-A-72360.

Rosenzweig, Daphne Lange. 2003. 'Continuity and Creativity: Modern Jingdezhen Ceramics'. *International Chinese Snuff Bottle Society Journal* 35(2): 26–33.

Scherzer, Georges F. 1900 [1983]. 'A Trip to Ching-te-chen'. In *Ching-Te-Chen: Views of a Porcelain City*, ed. Robert Tichane. Painted Post, NY: New York State Institute for Glaze Research, 187–201.

Sheel, Kamal. 1989. *Peasant Society and Marxist Intellectuals in China: Fang Zhimin and the Origin of a Revolutionary Movement in the Xinjiang Region*. Princeton, NJ: Princeton University Press.

Shyrock, John Knight. 1920. 'Kingtehchen The Porcelain City'. *Asia*, November 1920, 997–1002.

Sims, David. 2013. 'China Widens Lead as World's Largest Manufacturer'. *Thomasnet News*, 14 March 2013. http://news.thomasnet.com/IMT/2013/03/14/china-widens-lead-as-worlds-largest-manufacturer/.

Tang Ying. 1743 [1983]. 'Twenty Illustrations of the Manufacture of Porcelain'. In *Ching-Te-Chen: Views of a Porcelain City*, ed. Robert Tichane. Painted Post, NY: New York State Institute for Glaze Research, 129–73.

Tate Modern Gallery. n.d. 'Interpretation Text'. *The Unilever Series: Ai Weiwei: Sunflower Seeds*. www.tate.org.uk/whats-on/tate-modern/exhibition/unilever-series-ai-weiwei/interpretation-text.

Terpstra, Karen and Zhu Gui Hong. 2001. 'Firing Methods of a Wood-Fired Jingdezhen Commercial Kiln'. *Ceramics Technical* 12: 25–9.

Tichane, Robert. 1983. *Ching-te-chen: Views of a Porcelain City*. Painted Post, NY: The New York State Institute for Glaze Research.

*Time Out Shanghai*. 2012. 'Weekend Break: Jingdezhen', 16 November 2012. www.timeoutshanghai.com/features/Travel-Weekend_breaks/8873/Weekend-break-Jingdezhen.html.

Tinari, Philip. 2010. 'Postures in Clay: The Vessels of Ai Weiwei'. In *Ai Weiwei: Dropping the Urn, Ceramic Works, 5000 BCE to 2010 CE*. Glenside, PA: Arcadia University Art Gallery, 30–47.

Tyler, Patrick. 1995. 'With Deng's Influence Waning, Privatizing of China's State Industries Stalls'. *The New York Times*, 18 June 1995. www.nytimes.com/1995/06/18/world/with-deng-s-influence-waning-privatizing-of-china-s-state-industries-stalls.html?pagewanted=all&src=pm.

Uhalley, Stephen. 1966. 'The Cultural Revolution and the Attack on the "Three Family Village"'. *The China Quarterly* 27: 149–61.

Walder, Andrew. 1989. 'Social Change in Post-Revolution China'. *Annual Review of Sociology* 15: 405–24.

Wang Jianguo and Jiang Nan. 2007. 'Conservation and Adaptive-Reuse of Historical Industrial Building in China in the Post-Industrial Era'. *Frontiers of Architecture and Civil Engineering in China* 1(4): 474–80.

Wang Jifu. 2001. *Strategic Challenge, Strategic Responses, and Strategies: Study of Chinese State Owned Enterprises*. Auburn, AL: Ph.D thesis, Auburn University.

Wang Qian. 2011. 'From Local Ceramics to High-Tech Goods Exporter'. *China Daily*, 9 December 2011. www.chinadaily.com.cn/cndy/2011-12/09/content_14236875.htm.

Wang Qian. 2012. 'Expo Revitalizes Age-Old Ceramics Industry'. *China Daily*, 22 August 2012. http://usa.chinadaily.com.cn/epaper/2012-08/22/content_15697125.htm.

Wang Qingzheng. 2002. *A Dictionary of Chinese Ceramics*. Singapore: Oriental Art Publications.

Wang Zongda and Yin Chengguo, eds. 1994. 現代景德鎮陶瓷經濟史 1949–1993 [*Modern Economic History of Jingdezhen 1949–1993*]. Beijing: Zhongguo Shuji Publications.

Weiderman, Ken. 2015. 'Porcelain Reborn: CR's Group Show Honors the Porcelain City'. *The North Coast Journal*, 5 March 2015. www.northcoastjournal.com/humboldt/porcelain-reborn/Content?oid=2835844.

Whyte, Martin King. 1991. 'Urban Life in the People's Republic'. In *The Cambridge History of China* 15, eds. Roderick MacFarquhar and John King Fairbank. Cambridge: Cambridge University Press, 682–742.

Woetzel, Jonathan. 2008. 'Reassessing China's State-Owned Enterprises'. *Forbes*, 8 July 2008. www.forbes.com/2008/07/08/china-enterprises-state-lead-cx_jrw_0708mckinsey.html.

Wolfington, Vincent and Mark Wolfington. 2012. 'The Hospitality Talent Gap', *China Business Review*, 1 January 2012. www.chinabusinessreview.com/the-hospitality-talent-gap/.

Wood, Nigel. 2000. 'Plate Tectonics and Chinese Ceramics: New Insights into the Origins and Distribution of China's Ceramic Raw Material'. *Taoci* 1: 15–24.

Wright, Stanley. 1920. *Kiangsi Native Trade and Its Taxation*. Shanghai: n.p.

Wu, Anna. 2011. 'Jingdezhen Today'. *Arts of Asia* 41(6): 89–96.

Wu Xiaohong, Zhang Chi, Paul Goldbert, David Cohen, Pan Yan, Trina Arpin and Ofer Bar-Yosef. 2012. 'Early Pottery at 20,000 Years Ago in Xianrendong Cave, China'. *Science* 336: 1696–700.

Xiao Gongduo and Guo Guo'an. 2008. 'Jingdezhen Rising: The Return of a Porcelain Major'. *Asian Ceramics* 2: 18–26.

Xinhua News Agency. 2004. 'Important Discoveries in Jingdezhen'. *Xinhua* [New China] News Agency, 28 January 2004.

Xiong Xuehui. 2012. '景德镇十大瓷厂覆没' ['Jingdezhen's Ten Big Porcelain Factories Go Under.'] 中國經營報 [*China Business News*], 18 May 2012. n.p.

Yang Wanli. 2010. 'Porcelain Capital Hopes to Rekindle Age of Beauty'. *China Daily*, 13 October 2010, updated 10 April 2012. http:\\europe.chinadaily.com.cn/china/2012-04/10/content_15008824.

Ying Zilin, ed. 1988. 日用瓷工藝 [*The Art of Daily Use Porcelain*]. Nanchang, Jiangxi: Nanchang Linyun Publishers.

Zhang, Hanqin Qiu, Ray Pine and Terry Lam. 2005. *Tourism and Hotel Development in China: From Political to Economic Success*. Binghamton, NY: Haworth Hospitality Press.

Zhang, Michael. 2006. 'The Social Marginalization of Workers in China's State-Owned Enterprises'. *Social Research* 73(1): 159–84.

Zhou Jianshu, ed. 1989. 景德鎮史話 [*Speaking of Jingdezhen's History*]. Shanghai: People's Publisher.

# Index

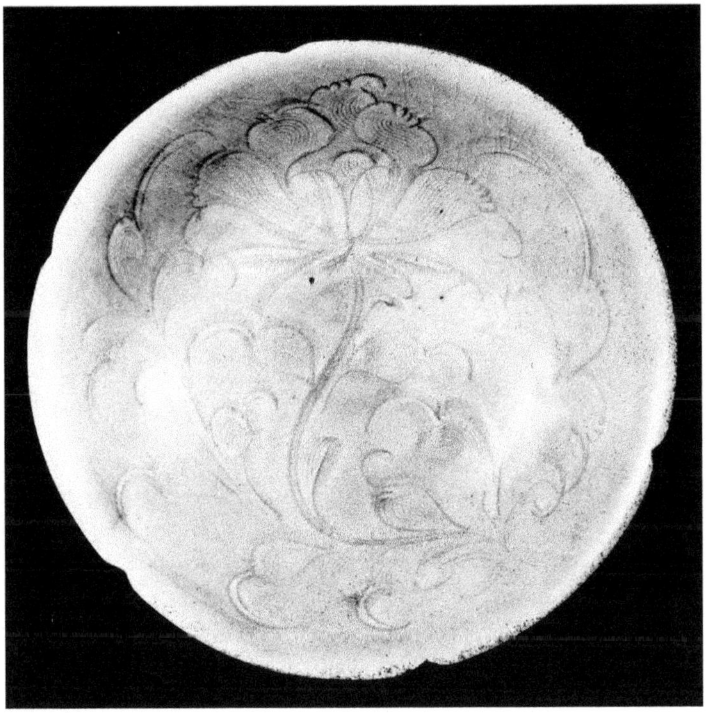

**Plate 1** A Song dynasty *qingbai* bowl from Jingdezhen dating to the twelfth century. 5.2 × 19.3 × 19.3 cm. The Dr. Paul Singer Collection of Chinese Art of the Arthur M. Sackler Gallery, Smithsonian Institution; a joint gift of the Arthur M. Sackler Foundation, Paul Singer, the AMS Foundation for the Arts, Sciences, and Humanities, and the Children of Arthur M. Sackler. RLS1997.48.1181.

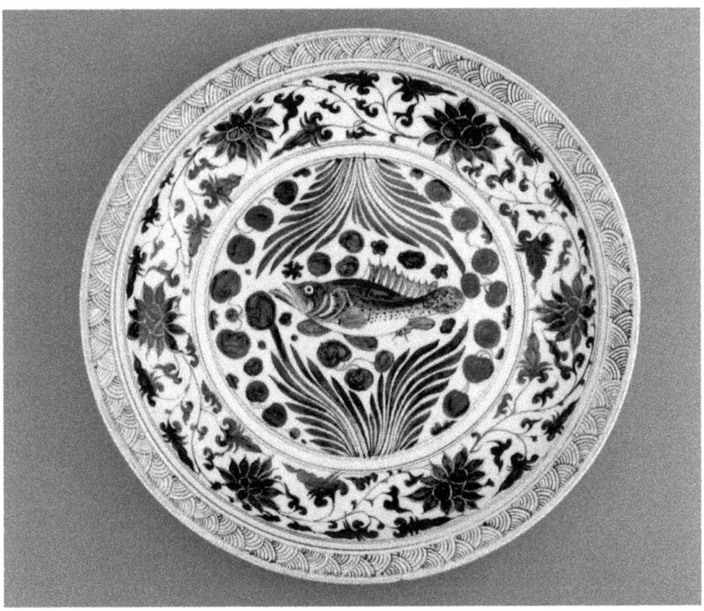

**Plate 2** A Yuan dynasty blue-and-white charger (porcelain decorated with underglaze cobalt pigment) dating to the mid-fourteenth century. Freer Gallery of Art, Smithsonian Institution, Washington, DC: Purchase. F1971.3.

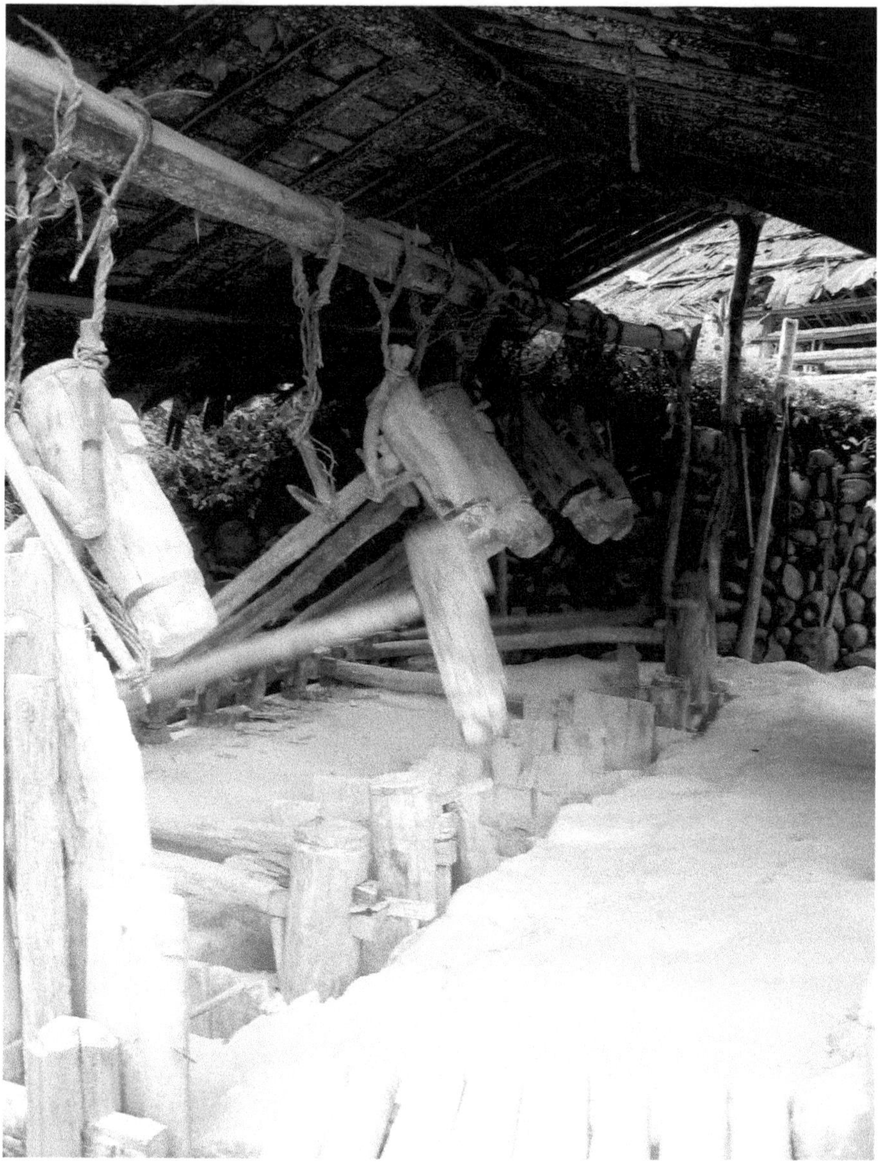

**Plate 3** A china stone or petunse crusher in the countryside near Jingdezhen. The hammers are powered by a water wheel turned by a stream. Production of 'little white bricks' (*baidunzi*) occurs during the spring, summer, and autumn, when water is abundant. Photograph by Maris Gillette, 11 September 2004.

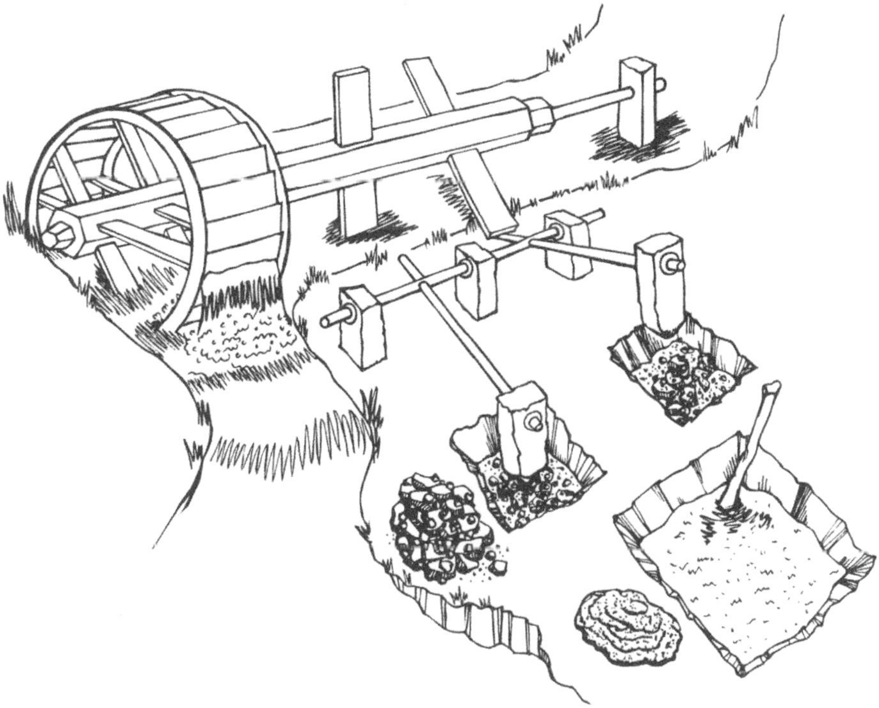

**Plate 4** This sketch depicts china stone crushers and a levigation pit. After the hammers pulverize the stone, workers transfer the powder to pits of water, where the impurities settle to the bottom and the fine china stone powder remains suspended. Workers skim off the slurry and allow it to dry partially before shaping it into small bricks. Potters later use the bricks to make clay. Drawing by Diana Tung, November 2015.

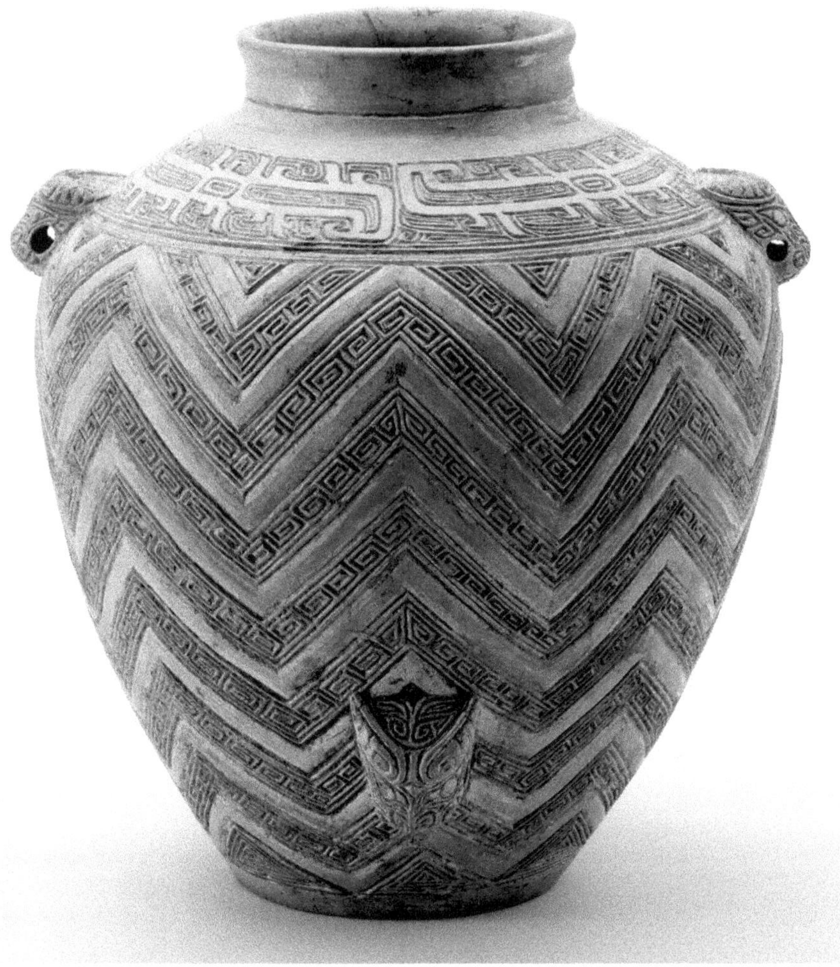

**Plate 5** A carved Shang dynasty urn from Henan Province, dating to the late thirteenth or the early twelfth century BCE. Freer Gallery of Art, Smithsonian Institution, Washington, DC: Purchase. F1939.42.

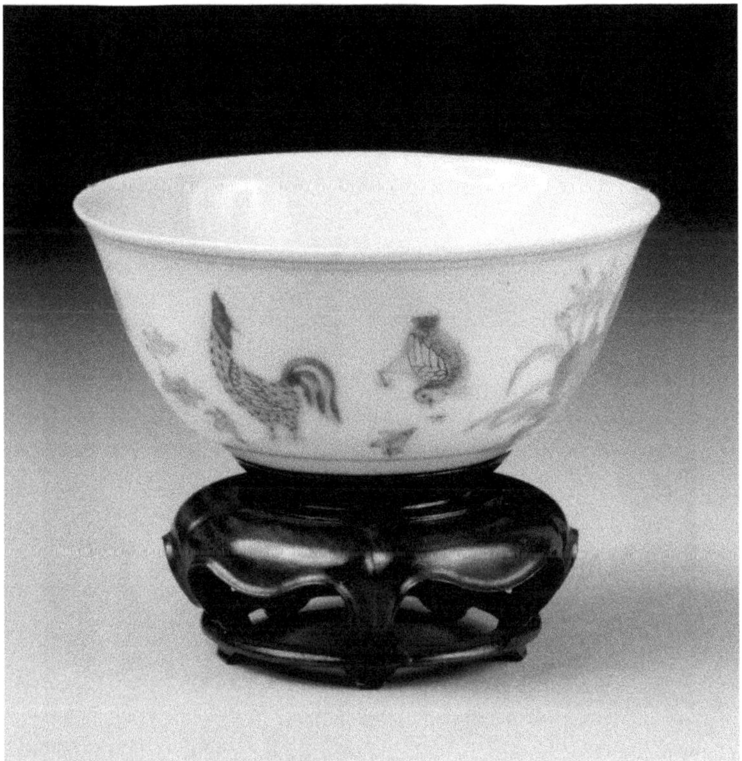

**Plate 6** This *doucai* or 'joined colours' chicken cup from the Chenghua period (1465–1487) of the Ming dynasty is one of only 17 still in existence. © Victoria and Albert Museum, London. C.1–1960.

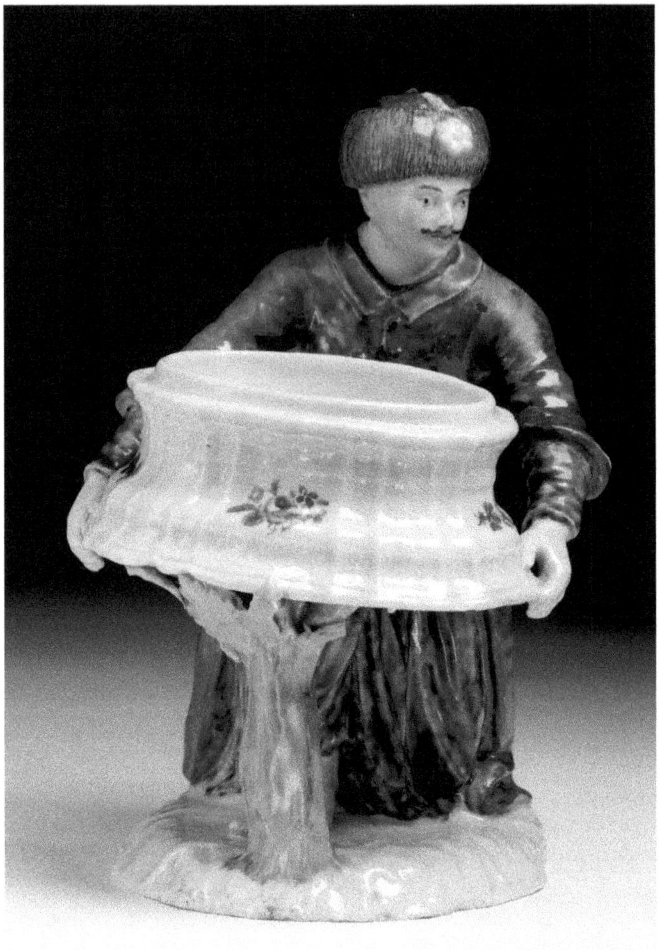

**Plate 7** This porcelain salt dish, decorated with overglaze enamels and gilt, was made in Jingdezhen around 1760 for export to Europe. Jingdezhen potters appear to have copied a figure modelled by Johann Friedrich Eberlein for Meissen Porcelain circa 1746. © Victoria and Albert Museum, London. C. 13–1951.

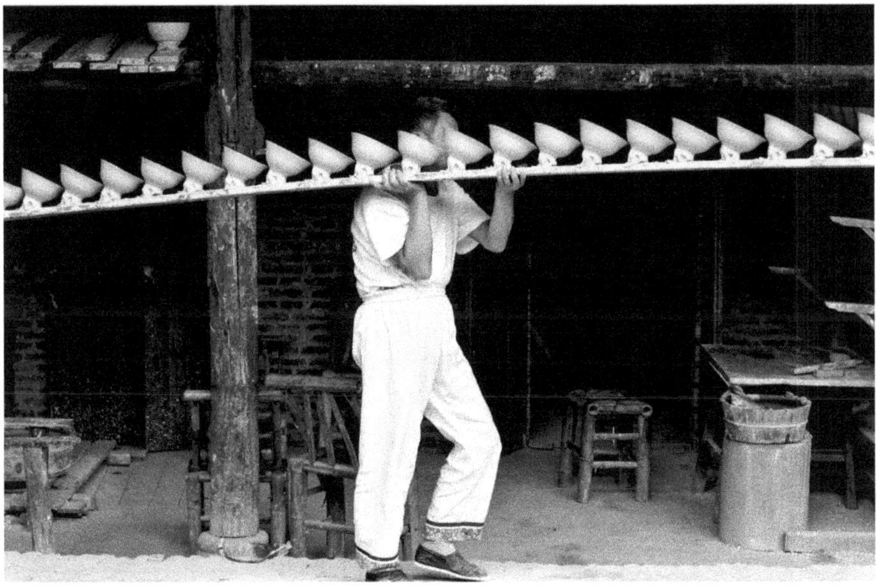

**Plate 8** François Philipp photographed this porcelain worker carrying a board of greenware at the Ancient Kiln Folk Customs Museum in 2013. This technology was ubiquitous through the late 1950s, and remained in use for decades longer at some state and collective factories. © François Philipp, 7 October 2013. Creative Commons License. www.flickr.com/photos/frans16611/11170258386/sizes/o/.

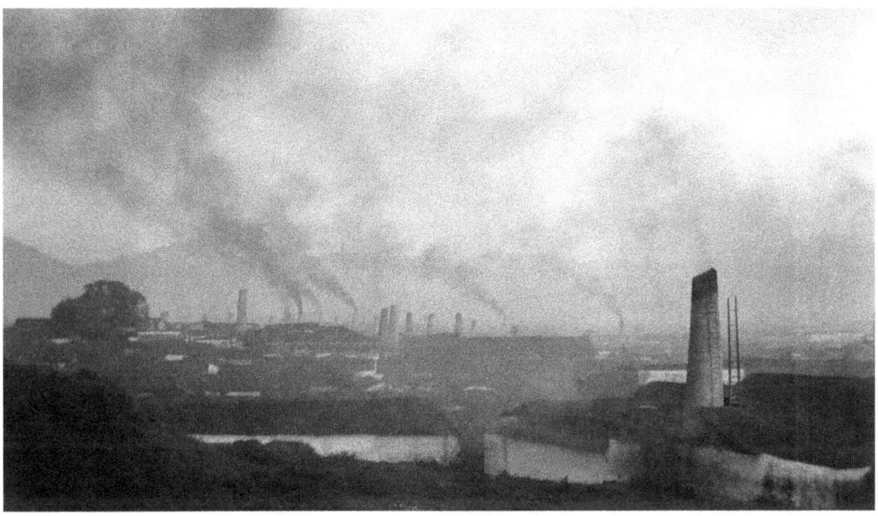

**Plate 9** Frank B. Lenz, a Christian missionary living in the provincial capital of Nanchang, published this 'bird's-eye view' of Jingdezhen in the *National Geographic Magazine* in November 1920. Lenz visited in 1919 or 1920 and was surprised to learn that the city had no electricity, telephones, or newspapers. In his article about 'the world's ancient porcelain center', Lenz wrote that 'long hours, poor food, no rest days, and unsanitary living conditions cause a great deal of dissatisfaction among the laborers'. Photograph by Frank B. Lenz/National Geographic Creative, June 1919 or 1920.

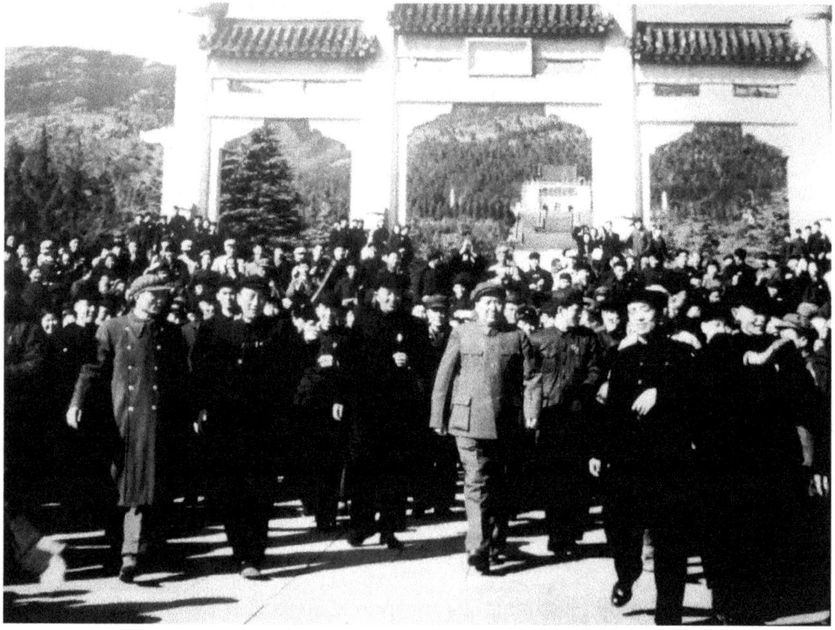

**Plate 10** Almost everyone visible in this photograph from February 1953 is wearing a Mao suit, also called a Sun Yat-sen suit. The photograph shows Chinese Communist Party leader Mao Zedong and others visiting the tomb of Sun Yat-sen in Nanjing. Creative Commons ShareAlike License, 23 February 1953.

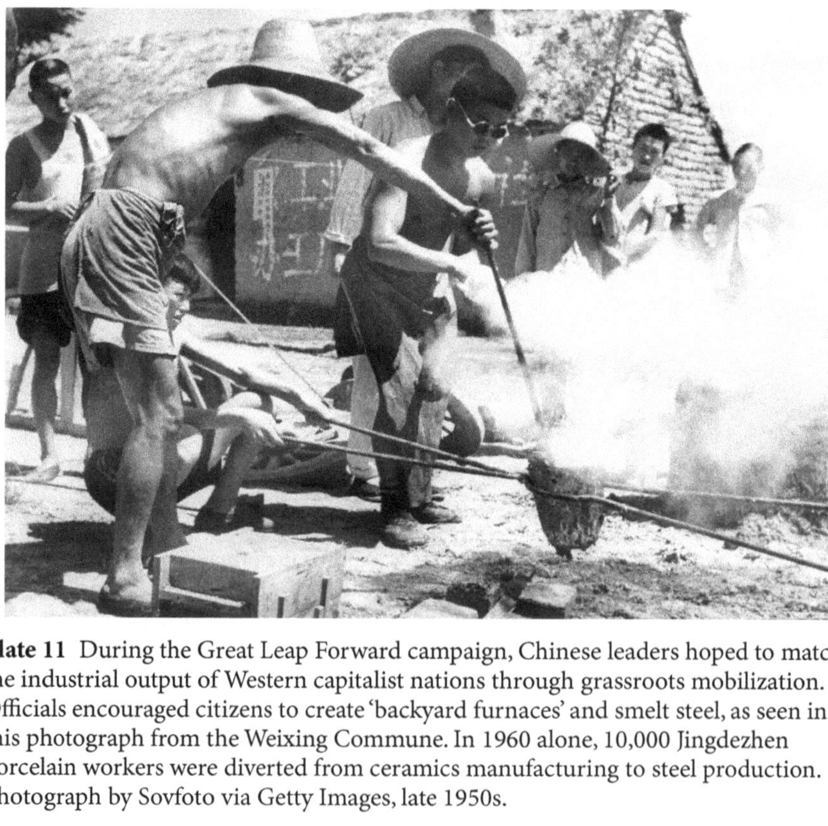

**Plate 11** During the Great Leap Forward campaign, Chinese leaders hoped to match the industrial output of Western capitalist nations through grassroots mobilization. Officials encouraged citizens to create 'backyard furnaces' and smelt steel, as seen in this photograph from the Weixing Commune. In 1960 alone, 10,000 Jingdezhen porcelain workers were diverted from ceramics manufacturing to steel production. Photograph by Sovfoto via Getty Images, late 1950s.

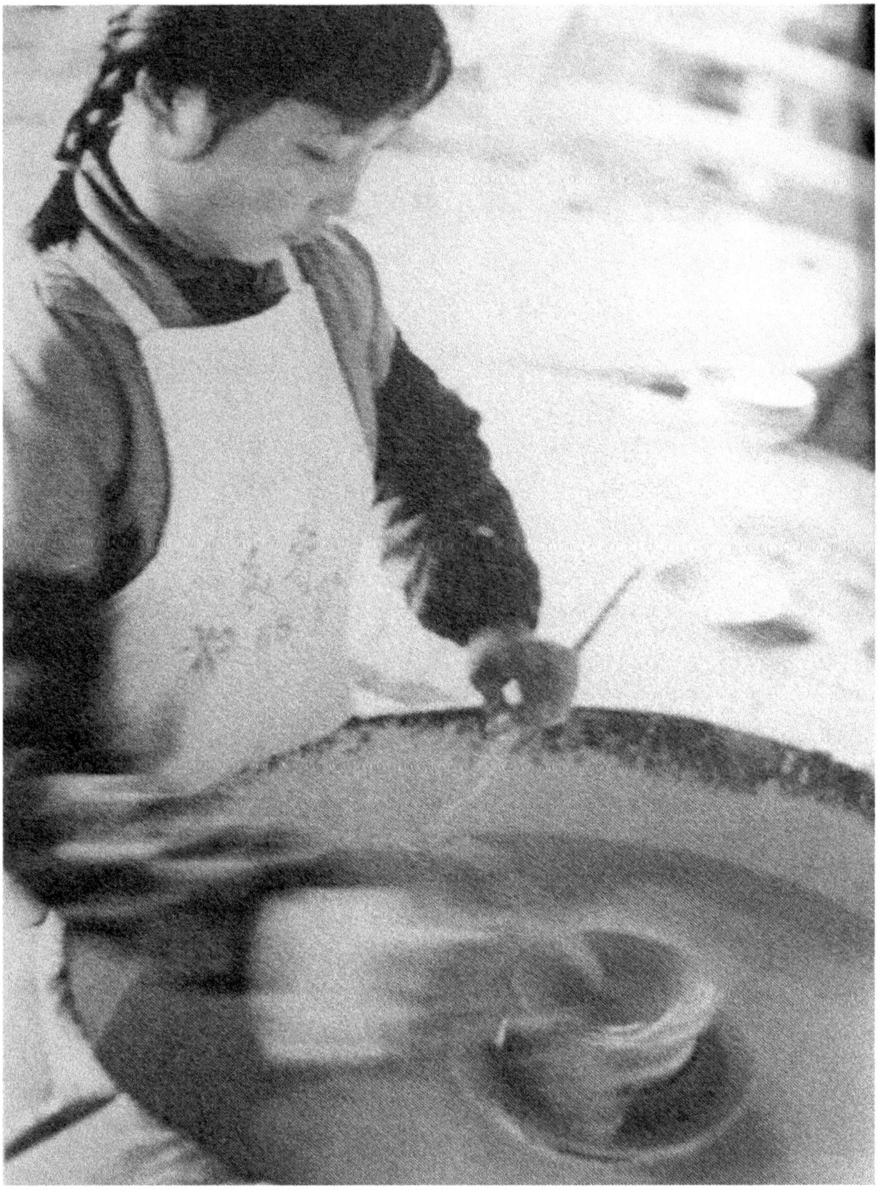

**Plate 12** Frank Cosentino of Boehm Porcelain admired the foot-pedalled glazing machines that he saw at the For the People Porcelain Factory in 1974. He described how the operator pushed her pedal to make liquid glaze shoot up from the tub and coat the ware. Photograph by Maurice Eyeington, December 1974.

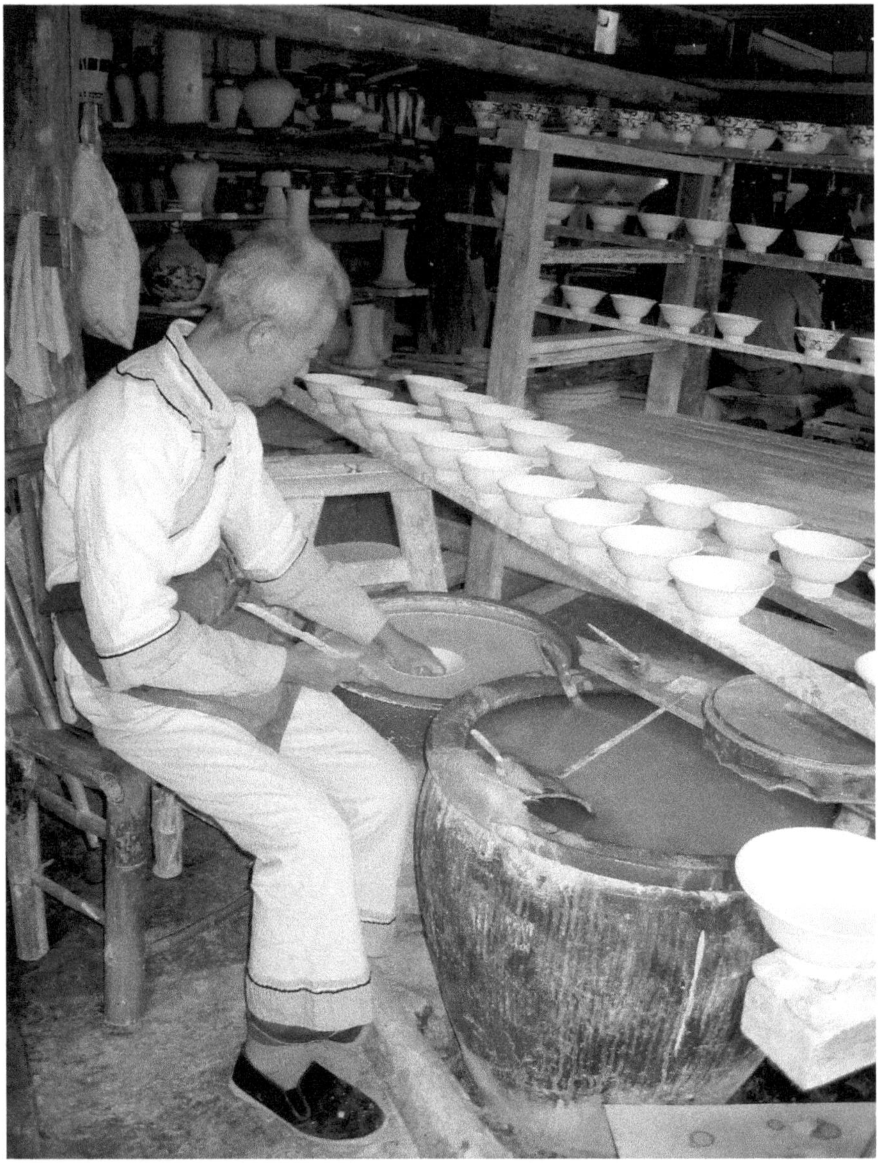

**Plate 13** This employee at the Ancient Kiln Folk Customs Museum glazed bowls by dipping. He and several other members of the staff came out of retirement to demonstrate traditional production methods for visitors. Photograph by Maris Gillette, 10 September 2004.

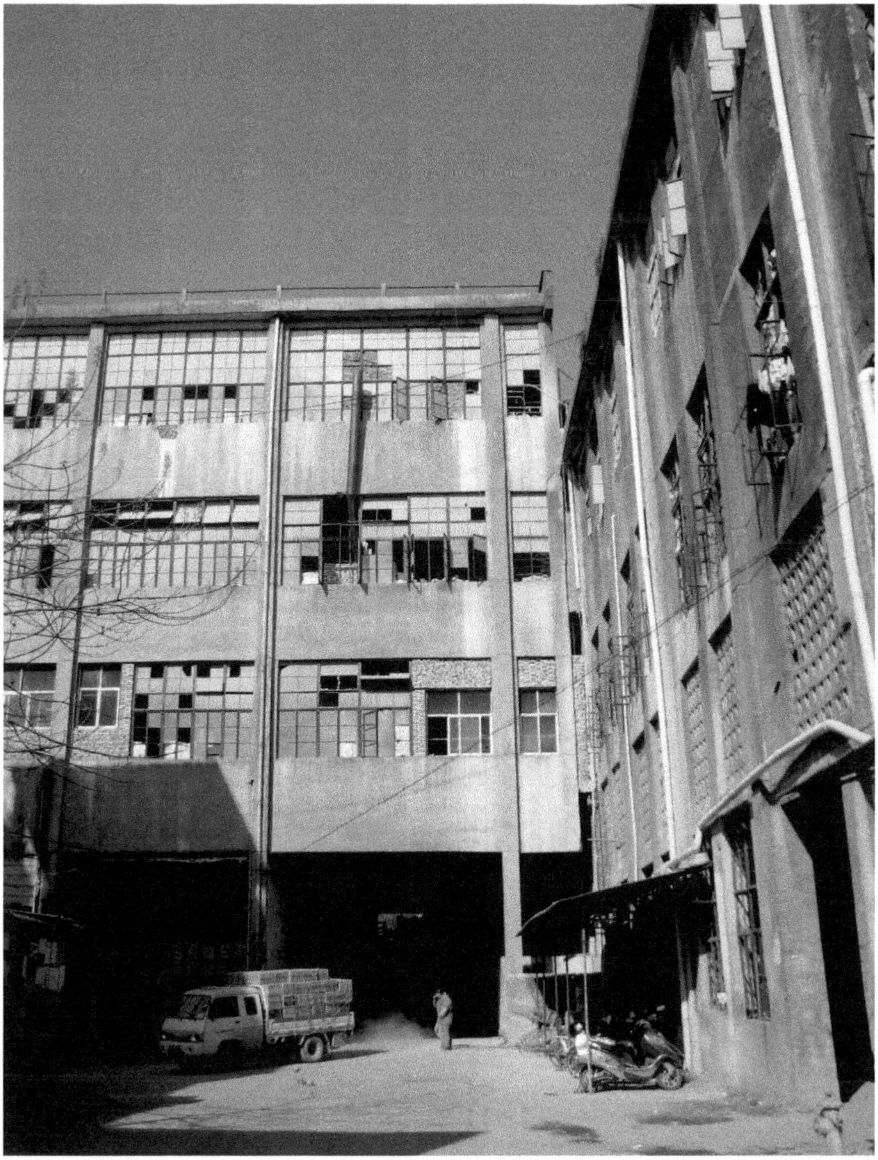

**Plate 14** The People's Porcelain Factory, January 2009. This state-owned factory won numerous ceramics prizes during the 1970s and 1980s, and was visited by national leaders such as Deng Xiaoping, who was China's paramount leader from 1978 to 1992. The People's Porcelain Factory went bankrupt in 1996, a casualty of the central government's decision to privatize the state sector. In 2009, private entrepreneurs rented some sections of the factory to produce rice bowls, while other areas were abandoned. Photograph by Diana Tung, 10 January 2009.

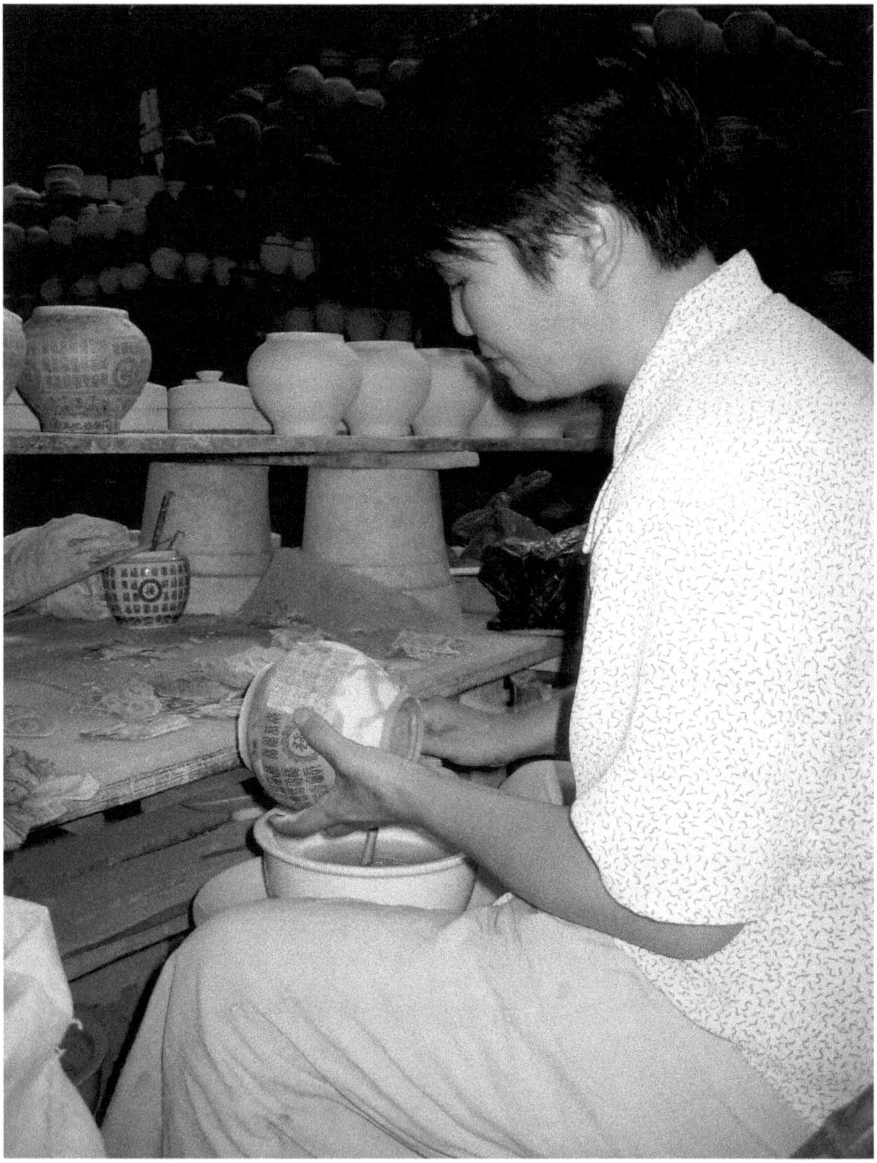

**Plate 15** After the state and collective porcelain enterprises collapsed, porcelain workers over 35 had a difficult time finding jobs. Older women with experience as surface decorators could apply transfers or decals for private entrepreneurs, as this woman is doing. The work was temporary and paid by the piece. Photograph by Maris Gillette, 19 September 2004.

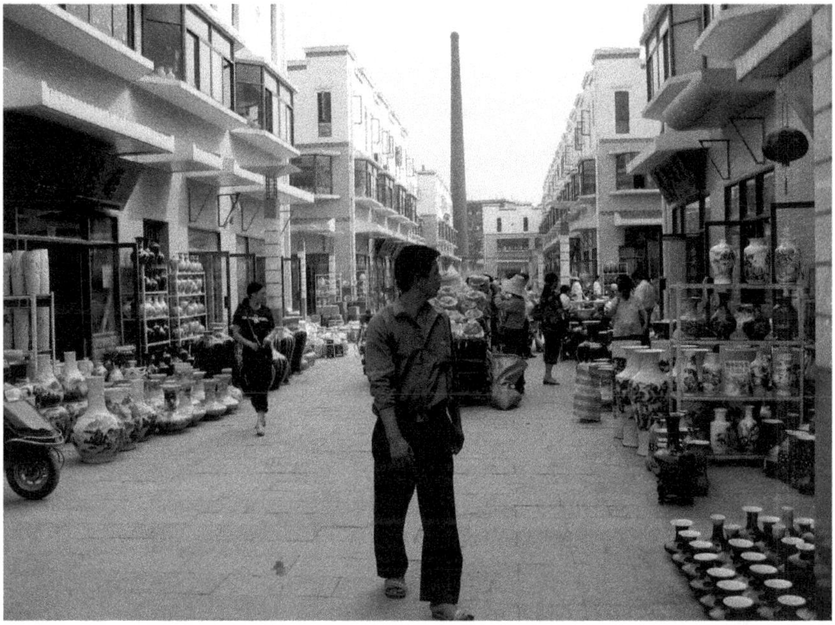

**Plate 16** The International Porcelain Market, shown here in September 2004, was built on part of the East Wind Porcelain Factory site after the former state enterprise was sold in 2002. A private developer collaborated with city officials to build storefronts, many of which were completed by the end of 2003. Photograph by Maris Gillette, 19 September 2004.

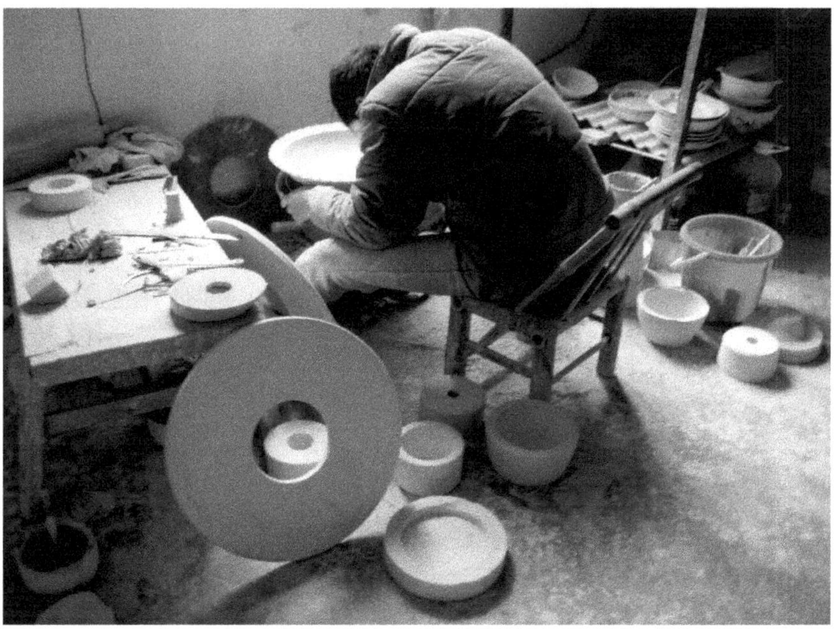

**Plate 17** When I met 24-year-old Yang Yajun in 2005, he was a private entrepreneur who made high-quality replicas of Yuan dynasty porcelain. Yang built forms by hand-pressing clay into moulds that he made from plaster of Paris. In this photograph Yang is carving surface decorations on a greenware charger. Photograph by Diana Tung, 8 January 2009.

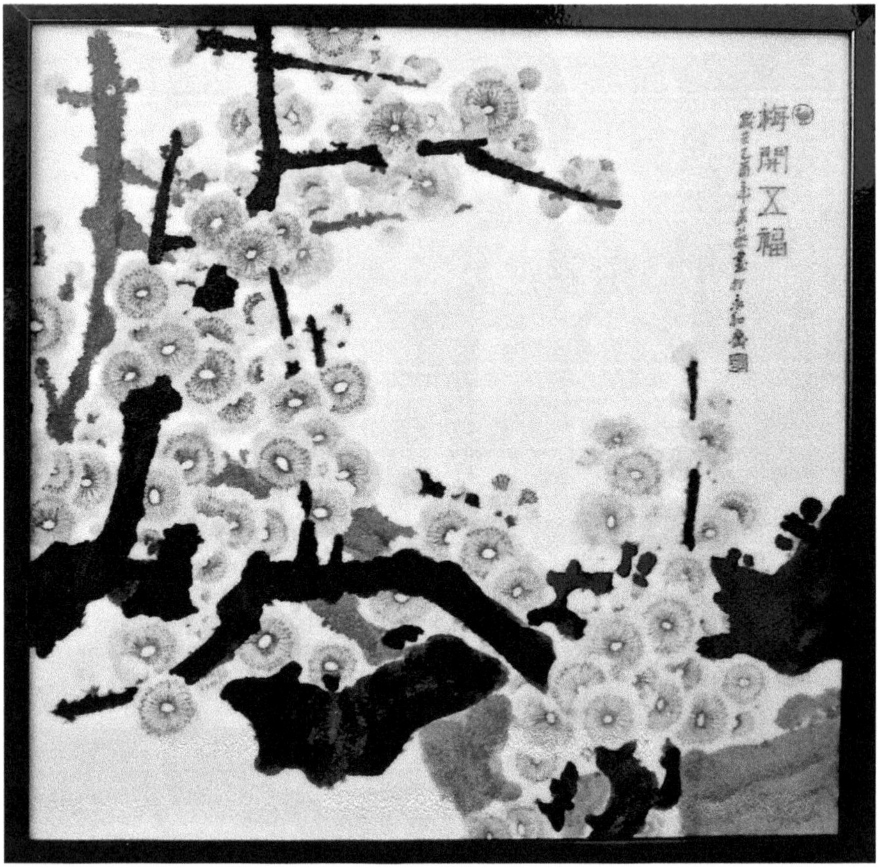


**Plate 18** Ceramic artist Wu Yan painted this cherry-blossom tile in October 2005. Wu was one of a select group of cadres' children who learned handicraft techniques from older masters in the late 1970s. She left her job in a state porcelain factory to become a private porcelain entrepreneur in 1996. In the 2000s, Wu Yan switched from selling historic replicas to contemporary art. Photograph by Mohsen Ghodsi, August 2015.

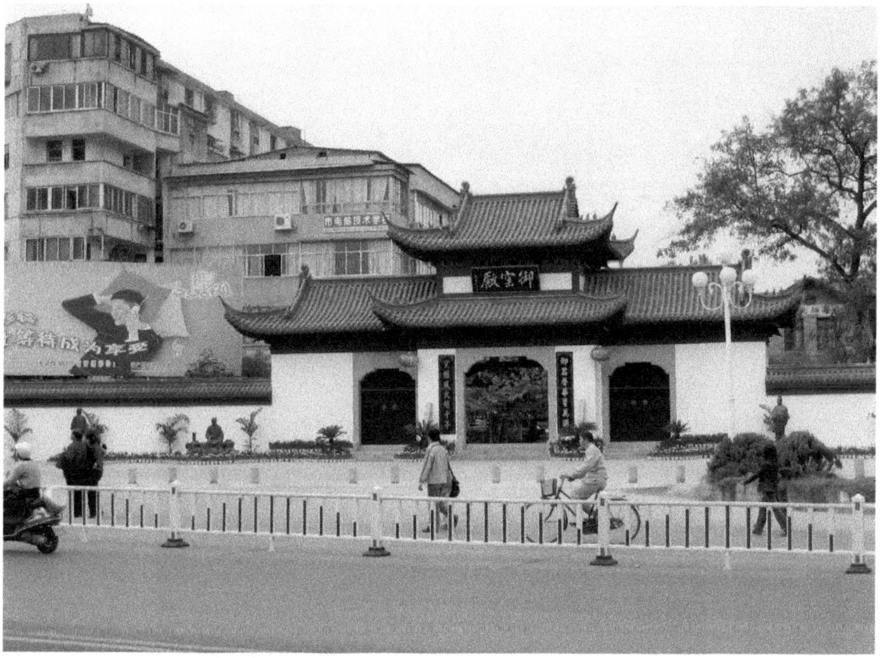

**Plate 19** Entrance to imperial kiln excavation site, October 2005. Municipal officials hosted a celebration to commemorate excavation of the imperial manufactory on 31 December 2003. Eight months later, facilities at the site were still very simple. By the autumn of 2005, the city government had added signage, a fancy gate, public art, and lighting. Photograph by Maris Gillette, 25 October 2005.

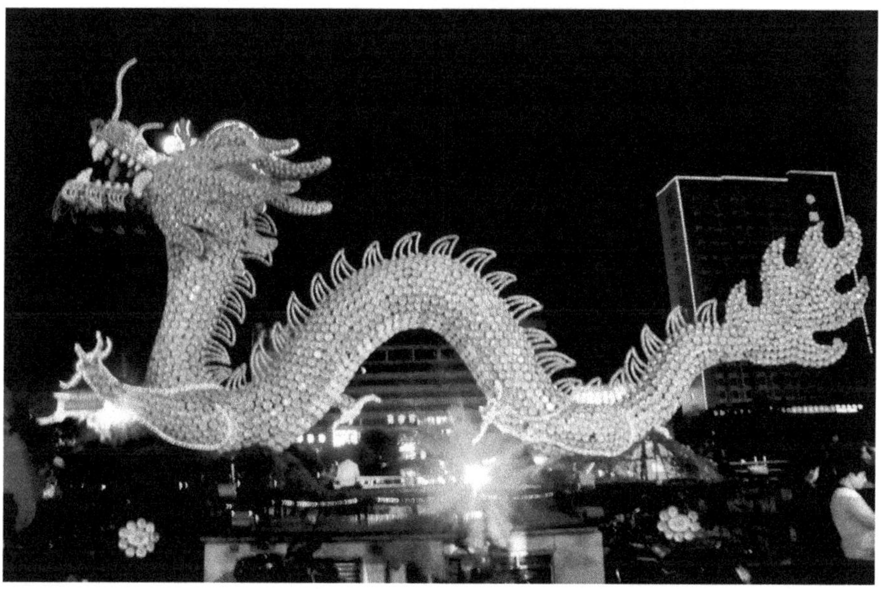

**Plate 20** This porcelain dragon, made from 40,000 porcelain rice bowls and cups, stood in the People's Square in 2012. City officials commissioned public art as part of their efforts to redevelop Jingdezhen and make the city attractive to tourists. Photograph by TPG/Getty images, 5 October 2012.

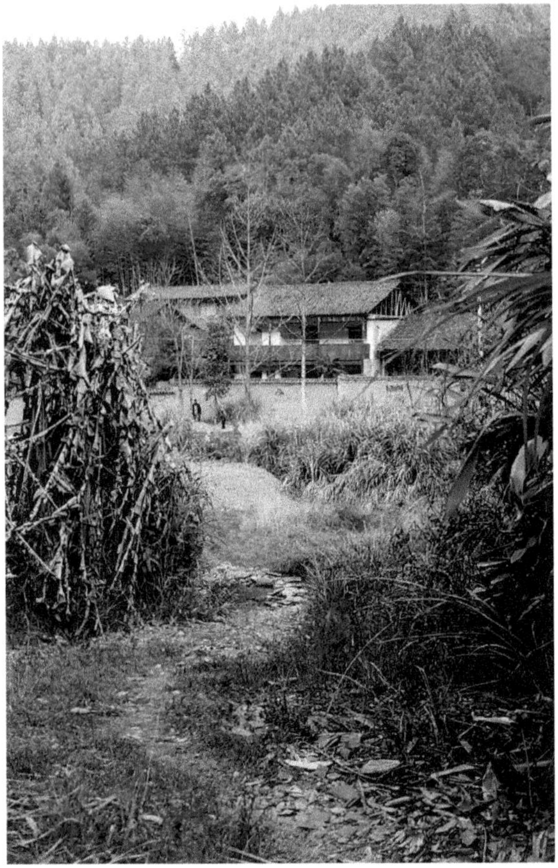

**Plate 21** The Sanbao International Ceramic Art Institute, shown here in 2003, was the first place in Jingdezhen to offer private artistic residencies. Li Jianshen founded Sanbao as an artists' colony in 1998, designing the site as a 'living tradition' that evoked the lifeway of a Chinese scholar and provided a comfortable work environment for contemporary artists. Photograph by Maris Gillette, 30 December 2003.

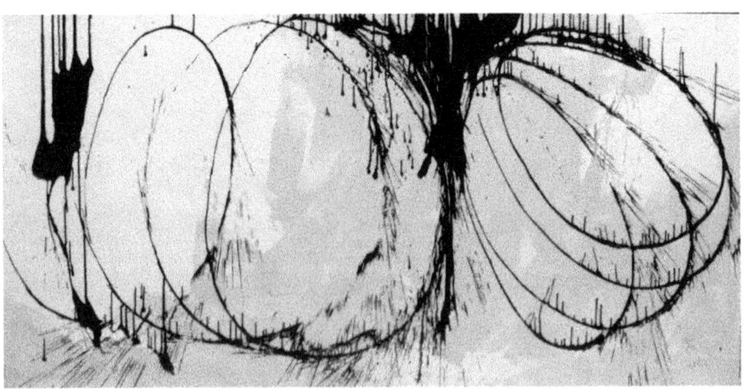

**Plate 22** This plaque by Norwegian artist Ole Lislerud is entitled www.Fortune.com and measures 2 metres × 1 metre × 1 centimetre. Lislerud began making large plaques during a residency at the Sanbao International Ceramic Art Institute in 2002, and continued to experiment with the plaques for the next decade. Photograph courtesy of Ole Lislerud, 2007.